THEORY AND RESEARCH IN LEARNING DISABILITIES

EDITED BY

J. P. DAS, R. F. MULCAHY, AND A. E. WALL

University of Alberta, Edmonton, Alberta, Canada

PLENUM PRESS · NEW YORK AND LONDON

Library of Congress Cataloging in Publication Data

International Conference on Theory and Research in Learning Disability (1980:
  Edmonton, Alta.)
Theory and research in learning disabilities
  "Proceedings of an International Conference on Theory and Research in Learning
Disability, held November 13–14, 1980, in Edmonton, Alberta, Canada"—Verso t.p.
  Includes bibliographical references and index.
  1. Learning disabilities—Congresses. I. Das, J. P. (Jagannath Prasad) II. Mulcahy,
R. F., Date –   . III. Wall, A. E. IV. Title.
LC4704.I57 1980   371.9   82-12219
ISBN 0-306-41112-1

Proceedings of an International Conference on Theory and
Research in Learning Disability, held November 13–14, 1980, in
Edmonton, Alberta, Canada

© 1982 Plenum Press, New York
A Division of Plenum Publishing Corporation
233 Spring Street, New York, N.Y. 10013

Printed in the United States of America

PREFACE

Although a number of books have appeared on learning disabilities, we feel that the present book has two distinct features which are not found in most others. It is multidisciplinary and it combines theory with practice.

A group of researchers from the disciplines of Psychology including medical psychology and information processing, Reading, Special Education and Physical Education interacted with each other before and after their presentations in a conference (November 1980, at the University of Alberta, Edmonton, Canada), and then wrote their chapters for this book. We hope that their chapters were enriched by the discussions and arguments which happened in formal and informal settings during the authors' stay in Edmonton. Dr. Leong could not attend the conference, but was asked to write the introductory chapter for the book. The contributors to this volume have been involved with basic research as well as with clinical and educational research in learning disabilities. Some of them have a theoretical rather than a practical interest. However, a serious interest in a handicapping condition perhaps compels one to consider its amelioration no matter how 'pure' is the researcher. It is because of such a feeling that those who do basic research have suggested procedures for applying their findings. The result is a balanced product, entailing theory and practice.

Several colleagues and graduate students at the University of Alberta's Centre for the Study of Mental Retardation, Faculties of Education and Physical Education and the Department of Psychology, have given invaluable assistance in organizing the conference; we wish to recognize all of them. We wish to thank Dr. Judy Lupart for coordinating the conference arrangements. In preparation of the book, Anne Price has helped us in editorial work, acting as a consultant and copy editor for the manuscripts. We would like to thank Barbara McGowan for typing most of the chapters, often twice and making camera-ready copies for the publishers.

v

We greatly appreciate the efforts of our authors for their scholarly contributions and certainly for putting up with editorial suggestions and changes. We do not think that the chapters cover the entire range of theories and research in the field of learning disabilities, but hope that whatever ground they cover, they do it in a readable manner.

THE CONFERENCE WAS FUNDED BY A THREE YEAR GRANT FROM IBM, CANADA, AS PART OF A PROJECT ON LEARNING DISABILITIES; we acknowledge the generous support of IBM. We also acknowledge the support of the Alumni Fund of the University of Alberta.

J. P. Das

R. F. Mulcahy

A. E. Wall

April, 1982

Edmonton, Canada

CONTENTS

PART I

Issues and Concepts

The three chapters provide an excellent introduction to concepts and issues in learning disabilities. Leong reviews definition, processes, etiology, and training procedures in the growing field of learning disability. He discusses reading and comprehension at some length in trying to understand specific reading disability, and leans towards a neuropsychological explanation of dyslexia. Leong's chapter anticipates the debates and discussions that follow in the remainder of the book.

Barbara Keogh offers a state of the art statement on learning disabilities, and like a senior statesman in the field, she also offers guidelines for some desirable research. After attempting to describe the frightful confusion in diagnostic techniques used to separate learning disabled individuals from others (she lists the prevalence of some 1400 techniques, which is a reductionist's nightmare), she looks at the therapeutic and interventional practices. The latter are justly described as incoherent and atheoretical. She also touches on the inadequacy of definition of learning disability, particularly the exclusionary criteria in the definition which seem to be a general concern of many researchers in the field, and a concern shared by the authors of the third chapter in this section. Barbara Keogh recommends that we study the interaction between motivational and affective characteristics on the one hand, and cognitive competence on the other while examining the nature of learning disabled children. She is justified in taking all of us to task for having paid so little systematic attention to the non-cognitive aspects of learning disability.

In the third chapter, Hagin and her coworkers discuss the importance of definition of learning disability in clinical settings. They too point out that as a practitioner one could not benefit from the exclusionary clause in the official definition because children never arrive at the clinic with a learning problem which is independent of other handicapping conditions. Their bias is for a multi-axial definition of learning disability which should be modified to suit different purposes for which the clinician or

1

the researcher is considering a sample of learning disabled children.
However, it was gratifying to see that the diverse symptoms found in
learning disabled individuals could be brought together as difficulties in spatial and temporal organization. The last two concepts
are essentially identical with simultaneous and successive processing, which is introduced by Leong but is elaborated in a chapter in
the next section.

A basic question comes to one's mind at the end of these major
reviews of the concept of learning disability and trends in research.
Is learning disability so closely related to school achievement that
we cannot define it independently? In other words, if the term
'learning disability' is to be taken simply as the words connote,
that is, disability in learning, then why could not it be found
outside schooling? For instance, would a learning disability exist
in a community without schools or informal teaching of reading and
writing, such as a primitive community? If it does exist in such a
community, how can we recognize it?

PROMISING AREAS OF RESEARCH INTO LEARNING DISABILITIES WITH EMPHASIS ON READING DISABILITIES[1]

Che Kan Leong

Institute of Child Guidance and Development
University of Saskatchewan
Saskatoon, Canada

This chapter as the first in the series, will attempt to draw out questions and areas of needed research having direct relevance to a better understanding of learning disabilities. In order to make this task more manageable I will focus on reading difficulties and specific reading disabilities in children. The following promising areas of research will be explored: meaningfulness or otherwise of the concept of learning disabilities with reference to associated or dissociated academic skills; component processes of reading and reading disorders related to decoding and comprehending; neuropsychology of reading disorders; orthography differences and the role of language awareness. These areas are obviously not intended to be exhaustive.

MEANINGFULNESS OF "LEARNING DISABILITIES"

One question at the outset is the meaningfulness of the term "learning disabilities." As originally proposed (Kirk, 1963) the term was meant to denote those children with language disabilities, who are otherwise relatively free from other disorders. Subsequently, the 1967 National Advisory Committee on Handicapped Children

---

[1] The writing of this paper and several of the studies were supported in part by research grants from the University of Saskachewan President's Humanities and Social Sciences Research Fund and from the June L. Orton Trust of the Orton Dyslexia Society. I want to thank Dirk J. Bakker, Byron P. Rourke and Otfried Spreen for interactions at various times. Any shortcoming is necessarily my own.

(NACHC) definition formed the basis of the U.S. Education for All
Handicapped Children Act or Public Law 94-142 (U.S. Office of
Education, 1976). Both the spirit and the letter of PL 94-142 are
praiseworthy in that the intent is to provide more and better services of all handicapped children including those with learning
disabilities. However, the concept and definition of this term have
broadened over the years and are liable to be too inclusive. This
broadening of the term led Cruickshank (1978) to argue for the
definition of learning disabilities within a developmental framework
of acquisition of academic skills and social, emotional growth. He
suggested that learning disabilities resulting from perceptual and
linguistic processing deficits should be redefined without any
regard for etiology, age, and level of intellectual functioning.
In early 1981 the National Joint Committee for Learning Disabilities
(NJCLD) representing a number of professional organizations proposed
the following definition for learning disabilities:

> Learning disabilities is a generic term that refers to a
> heterogeneous group of disorders manifested by significant
> difficulties in the acquisition and use of listening,
> speaking, reading, writing, reasoning or mathematical
> abilities. These disorders are intrinsic to the individ-
> ual and presumed to be due to central nervous system
> dysfunction.

> Even though a learning disability may occur concomitantly
> with other handicapping conditions (e.g., sensory impair-
> ment, mental retardation, social and emotional disturbance)
> or environemtnal influences (e.g., cultural differences,
> insufficient/inappropriate instruction, psychogenic factors)
> it is not the direct result of those conditions or
> influences. (Perspectives on Dyslexia, February, 1981.)

Both the Cruickshank suggestion and the NJCLD conjoint proposal
are an effort to ameliorate, if not correct, the existing exclusionary definition of learning disabilities and to provide for more and
better diagnostic and habilitative services. While one may disagree
with aspects of these proposals, one would readily agree with their
intent of providing more efficient and effective services for individuals with learning disabilities. The focus of the NJCHD definition is on altered processes of acquiring and using information and
on programing for individual variations. Norwithstanding these
developments, the argument presented by Barbara Keogh (this volume)
is a reasoned one. She suggests that learning disabilities research
suffers from a multiplicity of identification methods, sample heterogeneity, use of single variables rather than multivariate techniques and undue emphasis on inter- and intra-child variables almost
to the exclusion of extra-child or ecological influences. From her
ongoing UCLA Marker Variable Project, she discusses the research

needs and systematic research strategies to accommodate the above
shortcomings. I would agree with her transactional model and her
emphasis on a broad range of both within- and extra-child variables
analyzed stochastically.

Associated or Dissociated Academic Skills

Furthermore, the different achievement skills and their disorders likely involve different, though related, cognitive processes. Take listening and speaking as an example. Even though both
skills somehow employ the same linguistic processes, "it does not
follow that the two activities are directly analogous" (Mattingly,
1972, pp. 134-135). Mattingly and his colleagues at the Haskins
Laboratories further suggest that listening and speaking are primary
linguistic activities while reading and writing are secondary linguistic activities and that speech cues and print symbols carry different levels of linguistic information. The language cues for
listening-speaking are largely phonetic; the reading-writing symbols
involve the more abstract phonological, syntactic and semantic representations. Within the secondary language activities, reading
words correctly and spelling words correctly and, by inference,
reading and spelling difficulties, should be two sides of the same
coin "if the reading process itself mimics the spelling process"
(Frith, 1980a, p. 515). That both reading and spelling are correlated and mediated by rules is not disputed (see Baron, Treiman,
Wilf, & Kellman, 1980). What needs to be determined is the degree
and nature of the relationship and how this affects children with
learning disabilities. Thus, Uta Frith (1978, 1980a; Frith & Frith,
1980) distinguishes between "mildly dyslexics" retarded in both
reading and spelling relative to general intelligence and
"unexpected poor spellers" who are above average in reading. From
an analysis of spelling errors of these different groups of spellers,
she suggests that reading and spelling could be dissociated as the
mildly dyslexics make inconsistent errors which are also not
phonetic whereas the unexpected poor readers could benefit from
knowledge of orthographic structure. Thus Nelson (1980; Nelson &
Warrington, 1974) speculates that specific spelling disability may
be due to inefficient or faulty access to semantic information while
phonemic analysis of words may be relatively intact. Seymour and
Porpodas (1980, p. 471) point out that spelling production is
"functionally distinct from reading" in that in poor spellers the
"permanent storage" of orthographic information is impaired.
Current research into cognitive processes in spelling (see Frith,
1980b) and into reading (see Downing & Leong, 1982, chapters 9 and
10) all emphasize the role of phonological coding and that speech-
based deficiencies could explain certain aspects of the related
activities of reading and spelling difficulties. As with reading
words correctly and accessing word meaning, the nature of the
spelling code and the spelling access route would need to be further

explored. As spelling and reading may be dissociated, so too would
arithmetic skills from other academic skills. As an example, the
processes by which children solve single-digit subtraction problems
vary from incrementing or decrementing depending on which heuristic
is faster (Woods, Resnick, & Groen, 1975). Problem solving with a
number system depends a great deal on the way the problem is detected, the need to plan ahead and to consider alternative goals
(Resnick & Glaser, 1976).

It is thus clear from the above discussion that the term
learning disabilities as presently defined, as suggested by
Cruickshank or as proposed by NJCLD, serves its purpose best in
terms of service delivery. Among other aspects, the NJCLD definition draws attention to altered processes of acquiring and using
information and to the need to provide for individual differences.
Nevertheless, for research and scholarly purposes it behooves the
researcher to define exactly the group under investigation and to
state clearly the factors considered in the selection of experimental subjects. The importance of demarcation and operational
definition was emphasized by Spreen (1976) in summing up a conference on the neuropsychology of learning disorders. Only in this way
can we avoid confusion and promote greater understanding. The need
to demonstrate "clearly defined processing difficulties" and to
study the manner in which these processing disabilities affect performance of school tasks is further argued well by Torgesen (this
volume). Torgesen has buttressed his argument with a five-step
research program aimed at identifying processing difficulties in
children with learning disabilities and at finding ways of presenting materials that facilitate learning in these children.

COMPONENT PROCESSES OF READING AND READING DISORDERS

Within the all-embracing concept of learning disabilities as
discussed above, this chapter will concentrate on reading and its
disorders. At the outset, I would like to distinguish between those
children with reading difficulties and those with specific reading
disabilities or developmental dyslexia. The former group is generally termed "poor" readers or even "less skilled" readers and is
estimated to constitute about 10 to 12 or even 15 percent of the
school population. Within this 10 to 12 percent of school children
who experience reading difficulties, there is a small distribution
"hump" at the low end of the poor reading continuum. This subgroup
of children with specific reading disabilities or developmental
dyslexia presents many and varied severe reading disorders which are
probably due to central nervous system dysfunction. This is likely
the group that NJCLD focuses attention on in its proposed definition.

Whichever group or subgroup of disabled readers that we discuss, it is clear that the Zeitgeist points to the importance of the
cognitive or information-processing approach in understanding these
children and their reading difficulties. An early paper by Calfee,
Chapman, and Venezky (1972) entitled "How a child needs to think to
learn to read" highlights the importance of reading as thinking. In
his ongoing research program Calfee and associates have discussed
subskills in reading (Calfee, 1975, 1976, 1977; Calfee & Drum, 1978)
and have emphasized separable reading processes in "microanalysis"
of details and "macroanalysis" operating on text processing (Calfee
& Spector, 1981). The emphasis of reading as a skill, as an integrated skill welding different subskills, is highlighted in the
paper by Downing (this volume) and skill learning, be it motor or
verbal, has a large cognitive component (Whiting & Brinker, this
volume). The cognitive approach to reading psychology is a major
tenet in the Downing and Leong (1982) volume on psychology of
reading.

In essence, research in reading and reading difficulties has
moved away from examining or comparing products in the form of test
scores into one of teasing out processes and individual differences.
To the extent that reading is a verbal process, understanding
reading and reading difficulties must come from unravelling the role
of verbal processing as a key component of reading and a source of
reading difficulties. The central question evolves around the role
of "bottom-up" and "top-down" subskills or in operational terms the
independence or interdependence of decoding and comprehension. Put
another way, the question becomes one of: Are decoding subskills
both necessary and sufficient for comprehension?

Interdependence of Decoding and Comprehending

The series of studies by Cromer and Wiener (Cromer, 1970;
Cromer & Wiener, 1966; Oaken, Wiener, & Cromer, 1971; Stiener,
Wiener, & Cromer, 1971) was an early attempt to determine the verbal
source of reading difficulties. Their general finding is that decoding (defined in terms of word identification), though necessary,
is not sufficient for reading comprehension (in terms of standardized tests) and that high-level textual organization would help
comprehension with some groups of poor readers. Despite some
methodological problems and the narrowness in defining decoding and
comprehension in their studies, Cromer and Wiener have brought to
the fore the different sources of verbal difficulties in reading and
the need to study contextual devices in reading materials. Their
approach is the forerunner of some elegant, contemporary studies of
decoding and comprehension in skilled and less skilled readers.

In their ongoing research program Perfetti and his associates
at Pittsburgh (Berger & Perfetti, 1977; Goldman, Hogaboam, Bell, &

Perfetti, 1980; Hogaboam & Perfetti, 1978; Perfetti, 1977; Perfetti, Finger, & Hogaboam, 1978; Perfetti & Goldman, 1976; Perfetti & Hogaboam, 1975; Perfetti & Lesgold, 1977) have examined the nature of fast, accurate decoding of different stimuli in young readers and the question of interdependence of decoding and comprehension. Very briefly, they found three sources of individual differences with less skilled young readers. First, less skilled readers show longer vocalization latencies to context-free printed words and pseudowords, while they do not differ from skilled readers in color and picture naming tasks. Second, less skilled readers show slower lexical access (determining if a word is a word or a nonword). This relates to their difficulties with encoding letter strings and not so much with a poor lexicon. Third, less skilled readers are less efficient with phonological coding so essential to reading comprehension, even though the surface, syntactic and interpretive levels also need to be integrated (Perfetti, 1977). Results similar to those of Perfetti and associates were obtained in our laboratory by Haines (1978) in several experiments dealing with vocalization latencies and lexical decisions by skilled and less skilled readers in grades 4, 6, and 8 with carefully controlled predictable words, unpredictable words, and pseudowords varying in length. Haines found evidence to suggest that word processing speed continues to develop to at least about grade 8 and that less skilled readers are slower in processing words and also commit more errors. This general verbal processing deficiency is also found in dyslexics as inefficiency in syntactic, cognitive-semantic and logical processing (Vellutino, 1979); as lexical encoding deficiency (Ellis & Miles, 1981) and as difficulty in translating visual items into speech code (Legein & Bouma, 1981).

The Pittsburgh group is careful in pointing out that their findings of the verbal sources of reading difficulties should be interpreted within the framework of meaningful organization of language units and general language comprehension. This broader interpretation will in some way reconcile the variant finding by Mason and her associates (Mason, 1978, 1980; Mason, Pilkington, & Brandau, 1981) that vocalization latency differences in relation to reading ability in adults are not necessarily limited to alphabetic materials as 'found by Perfetti and his colleagues with young readers. Mason et al. (1981) suggest that reading difficulties relate to the very rapid time needed to encode and locate visual and auditory information. They state:

If we assume that the perception of when and the perception of where are closely interdependent and possibly mediated by a common mechanism, it is reasonable to believe that difficulties in resolving temporal order and resolving spatial order will also be interdependent (p. 590).

This basic concept is well put by Stanovich (1980, p. 62):

> ... an alphabetic writing system serves as a powerful
> reinforcer during the initial stages of reading only
> if the child has the phonological skills to make use of
> the alphabetic principle.

The corollary of this is that those children who are "linguistically
aware" are also likely to read more and this in turn leads to automaticity (see LaBerge & Samuels, 1974) and rapid word recognition
ability which is a component of fluent reading. This corollary is
tested directly by Lesgold and Resnick (this volume) in their longitudinal study of multiple reading skills of cohorts of children in
different instructional settings through primary grades. Their path
analysis confirms the earlier and ongoing findings of the Pittsburgh
group that automatic, rapid word processing is an important component of reading comprehension and that slow or non-automatic reading
is symptomatic of comprehension difficulties.

Comprehension Processes

This brings me to reading comprehension--a large area that
linguists, philosophers and more recently psychologists and researchers in artificial intelligence have been grappling with.
Considerable progress has been made towards understanding of reading
comprehension in general because of rigorous research from these
related disciplines (see the recent, comprehensive volume on reading
comprehension by Spiro, Bruce, & Brewer, 1980, among others). Much
less is known of reading comprehension in disabled readers and for
that matter even less has been written and put into practice in
remedial reading comprehension instruction. This paucity in research, theory and practice is recognized by Vernon (1979) in her
discussion of "variability in reading retardation" and by Myklebust
(1978) in his plea for a "science of dyslexiology." Among other
things, he stated that:

> A deficiency in ability to gain meaning, especially in
> relation to abstract thought, is the most significant
> consequence of a handicap or a disability, including
> dyslexia (p. 5).

In some ways, the slower progress made in reading comprehension compared with what is known in lexical access can be traced to
the narrow conception of language in terms of conditioning, mediation theory and information theory prevalent to the 1950s or so.
Take as an example the influential book Language by Bloomfield
(1933). The Bloomfieldian model is strictly bottom-up with phonetically transcribed utterances as primary data. Bloomfield was not
unaware of the importance of meaning except that he was constrained

by his own behavioristic orientation and also by the lack of
scientific principles and techniques to study meaning. He acknowl-
edged this weakness in language study:

> In order to give a scientifically accurate definition
> of meaning for every form of a language we should have
> to have a scientifically accurate knowledge of every-
> thing in the speaker's world... We can define the
> meaning of a speech-form accurately when this meaning
> has to do with some matter of which we possess
> scientific knowledge (Bloomfield, 1933, p. 139).

The focus of language learning of the 1930s to the 1950s is on what
Miller (1974) terms the "association" paradigm. It is not until
around the mid 1960s that studies began to concentrate on what
people do when they produce or understand linguistic signals. This
changing emphasis "should lead us to look for procedures involved
in using language" (Miller, 1974, p. 404).

   This emphasis on processes or procedures comes from significant
converging work from linguists, cognitive psychologists and researchers in artificial intelligence. From a linguistic perspective,
Fillmore (1968, 1971) discusses verb-noun relationship and its
extension in his case grammar; Pike (1967) writes of role relationship and plot development in prose; Grimes (1975) and Halliday and
Hasan (1976) discuss discourse structure and text cohesion. From a
psychological perspective, Kintsch (1974) provides us with a comprehensive theory of text processing and the representation of knowledge; Rumelhart (1977a, 1977b) has developed a formalism based on
parallel, interactive processing of reading and has emphasized the
memory system in comprehension and Anderson (1977) has elucidated
schema theory derived from the work of Piaget (1926) and Bartlett
(1932), which can in turn be traced to philosophical antecedents by
Kant (1781). The artificial intelligence counterparts of the Piaget
and Bartlett schemata are the concepts of "frames" of Minsky (1975)
and the "scripts" of Schank and Abelson (1977). In computational,
psychological terminology, a schema consists of a well-organized
semantic network of variable slots which must be instantiated or
filled with values within variable constraints. For example, the
general theme of "dining in a restaurant" will have a "food schema"
which can be filled with particular instances of "menu," "selection
of food," "form of payment" and the instantiation of these slots
with different values varies according to the world view of the
writer/reader (see also Spiro, Bruce, & Brewer, 1980).

   Thus understanding of prose, particularly of short stories,
involves understanding of related slots to be filled by specific
goals or events of the protagonist in a particular story. There
are rules governing the relationships weaving the theme or focus,

the plot or group of actions, the setting and the resolution of the
goal of the story. This framework of story grammar structure is
well explained by Thorndyke (1977), as well as among other related
grammars. Story comprehension thus takes the form of the selection
of abstract representation (schema) and it is the high-level representation that helps to organize a story and to make it comprehen-
sible and recallable.

The necessarily succinct discussion above of current developments in reading comprehension helps to explain the paucity of research and application of this emerging work for the reading disabled. For one thing, research into reading disabilities and diag-
nostic-prescriptive teaching has been dominated for too long by
"perceptual" theory and training and the notion that "lower-down"
subskills must be mastered before "higher-up" processes can be
dealt with. As is clear from the discussion of interactive processes, reading processes must use simultaneously the constraints at the
featural, letter, letter cluster, lexical, syntactic and semantic
levels to interpret prose materials (see Rumelhart, 1977b;
Stanovich, 1980). For another thing, the emerging theory of reading
comprehension instruction within the schema-theoretic framework,
which can account for reading proficiency and reading breakdown, has
yet to be formalized and further tested (see Carr, 1981; Pearson &
Spiro, 1980).

Training to Comprehend

Some ongoing work offers promise for teaching "schemata" to
reading disabled children (Pflaum & Bryan, 1982). In several
studies Pflaum and Pascarella (1980) and Pascarella and Pflaum
(1981) field-tested their instructional program emphasizing error
detection and self-correction and integration of meaning from
context. They found that those reading disabled children with grade
two reading proficiency benefitted from such a systematic program
and that locus of control interacts with training with the internal-
ly controlled disabled readers achieving greater gains in reading
than the externally oriented disabled readers. In reviewing their
work, Pflaum and Bryan are careful to point out the appropriateness
of the program applies only to those with at least a second-grade
level reading proficiency and that no direct, causal relationship
between training and reading performance is yet established. What
is shown is that strategy learning

    ... may heighten learning disabled children's knowledge
    of sentence components, which might, in turn, increase
    their ability to predict context-appropriate words
    (Pflaum & Bryan, 1982, p. 41).

What is emerging from the burgeoning literature is that in reading disabilities strategic learning must be taught and must be taught explicitly. In prose learning, this includes analyzing components and links of components, detecting errors and self-correcting them, testing for hypotheses, inferencing and integrating information. To complement this aspect of helping children to monitor their comprehension, the structure and organization of text materials for readers must be carefully studied (see Markman, 1981). Markman suggested that texts for young children usually consist of unconnected sentences or short sentences and paragraphs and this attempt at "simplicity" and explicitness for beginning readers and poor readers militates against the constructive and integrative processes so essential to reading comprehension. Downing and Leong (1982) have discussed at some length the schema view of reading; the importance of different linguistic devices (e.g., connectives, contrasts, anaphoras, polysemies) that assist text cohesion; and the representation of meaning in prose. The volume edited by O'Neil (1978) on learning strategies emphasizes training for knowledge structure.

Such training must be many faceted. Learning to learn over and above learning content areas must be integrative (bringing together inter-related concepts), inferential (going beyond the information given), and performance-oriented (applicable to different situations). The many facets of training in prose learning include training in the use of mental imagery, self-questioning and monitoring, reorganization of incoming materials, elaboration training to link new with old knowledge, building of networks of important concepts (nodes) and learning their inter-relationships. Training strategies must be specified in detail and should not simply rely on the inductive power of learning disabled or reading disabled children as they are less inner-directed. The training pace must be specified and moderated. Feedback to and motivation of the learners are important considerations. As well, long-term effects of strategy learning must be assessed. The weight of evidence from the O'Neil volume, especially the work of Dansereau (1978), points to the need for a sound conceptual framework and methodology of training which should and must emphasize the locus of specific effects and not just generalized benefits, as is common with many traditional training programs. Much more needs to be done in providing detailed information on the actual procedures in training learning strategies.

NEUROPSYCHOLOGY OF READING DISORDERS

The explicit statement of the presumed central nervous system dysfunction of learning disabilities in the proposed NJCLD definition brings to the fore the role of neuropsychological

contributions. That there are neuropsychological approaches with
relevant data base to reading and reading disorders can no longer be
disputed. The influential volume edited by Knights and Bakker
(1976) has discussed theoretical approaches to and empirical studies
of learning disorders. The equally significant volume of Benton and
Pearl (1978) has focused on the appraisal of current advances in
theory and research of dyslexia. In addition, an expository account
by Farnham-Diggory (1978) and a recent paper by Patterson (1981) all
emphasize the need to study learning disabilities and reading disorders within a brain-behavior context. In this regard, Spreen
(1976) suggests that attempts at building a neuropsychology of
learning disorders should be "modest" and that such theoretical
models must encompass both "normal and abnormal" learners and must
add to our understanding of learning processes and remediation
practices. Such a balanced view is also voiced by the editors of
the prestigious work Education and the Brain (Chall & Mirsky, 1978).
Within this stricture there are at least two promising areas worthy
of further research: laterality and reading proficiency and subtyping of disabled readers.

Laterality Studies and Reading Disabilities

        The term laterality refers to the specialization and the recip-
rocal relationship of the left and right hemispheres of the brain
for different psychological functions. From the early findings in
the 1960s that the left hemisphere is specialized for processing
written and spoken language and that the right hemisphere is
specialized for visuospatial stimuli have derived significant
theoretical and empirical studies elucidating the nature and locus
of hemispheric functions. Space does not permit a thorough review
of the vast and diverse literature in this area. Only the salient
aspects can be delineated. From his different experiments using
dichotic digit and letter tasks, Leong (1976, 1976-1977) suggested
that dyslexics lag behind their age- and ability-matched controls in
functional cerebral development. This lag could mean, as proposed
by Bakker (1979), that the nature of cerebral asymmetry and reading
ability could be dependent on the phase of the learning-to-read
process and that amongst dyslexics their reading performance may
relate to left or right cerebral laterality.

        In a recent review article focusing on hemispheric asymmetry
from visual half-field studies of normal and poor readers, Young and
Ellis (1981) emphasize the need for methodological refinements such
as fixation control, ensuring similarity of stimuli and equivalence
of cognitive processes. They claim that only when the same strategies or processes are used by both normal and poor readers can we
infer asymmetrical hemispheric organization for the same functions.
Leong (1980a) took a slightly different position in reviewing both
dichotic listening and visual half-field studies bearing on

laterality and reading in children. Leong emphasizes the
laterality-reading relationship as complementary contributions of
the two hemispheres to different stages of reading, to lingual and
perceptual types of errors, to different task requirements and to
processing strategies used by readers. Thus by a vicarious route,
we are led back to the importance of task analysis, of teasing out
processes of learning and of remediating these processes as variously emphasized in this volume by Das, Snart, Mulcahy, Hagen, Lesgold
and Resnick, and Torgesen, among others.

Subgroups of Disabled Readers and Pattern Analysis

        Studies of subgroups or subtype of learning disabilities aim at
a more homogeneous grouping of these individuals for meaningful research and remediation. Two broad approaches are generally followed.
One approach is psychometric or performance-level related; the other
is neuropsychological or clinical-inferential. Good examples of the
psychometric approach include: the investigation by Myklebust,
Bannochie, and Killen (1971) into the ability patterns of 116
"moderate learning-disability" children and 112 "severe disability
children," each compared with their controls; the factor analyses of
Wechsler Pre-School and Primary Scale of Intelligence (WPPSI) subtest
scores of above-average and below-average seven-year-old readers by
Maxwell (1972a, 1972b), who found different factor structures for the
two groups.

        For clinical samples of children with developmental dyslexia,
studies of interest were those by Doehring (1968), Doehring and
Hoshko (1977), Naidoo (1972), Leong (1980b), Mattis and associates
(Mattis, French, & Rapin, 1975; Mattis, 1978), Rourke and associates
(Fisk & Rourke, 1979; Petrauskas & Rourke, 1979; Rourke, 1978;
Rourke & Finlayson, 1978; Rourke & Gates, 1981; Rourke & Strang,
1978). Doehring (1968) found from his multivariate analyses of a
large number of sensory-motor, perceptual and verbal measures
administered to dyslexic boys and their controls that the dyslexics
displayed an "asymmetry of talents" and also deficits in "sequential
processing." Naidoo carried out a taxonomic analysis of the structure of performance on perceptual-cognitive tasks in dyslexic boys.
Leong predicated his studies of dyslexics and other poor readers on
the Luria (1966) paradigm of simultaneous and successive synthesis
(see also Das, Kirby, & Jarman, 1979; Das, Snart, & Mulcahy, this
volume) and found through different models of factor analyses differential factor patterns of the disabled readers. The Doehring,
Naidoo, and Leong studies are essentially studies of patterns of
impairment of disabled readers compared with their controls. This
approach is further emphasized by Rourke and Finlayson (1978) and
Rourke and Strang (1978) in teasing out patterns of neuropsychological abilities and deficiencies in relation to reciprocal
cerebral functions.

In a slightly different way, Doehring and Hoshko (1977) used
the Q-technique of factor analysis to study children with "reading
problems." These children were differentiated into those characterized by: (a) slow oral word reading, (b) slow auditory-visual
letter association, and (c) slow auditory-visual association of
words and syllables. Extending the Q-technique to 133 seven- and
eight-year-old "retarded" readers compared with 27 controls on 20
neuropsychological tests selected from over 40 such tests,
Petrauskas and Rourke (1979) found three subtypes of retarded
readers accounting for 50 percent of the subjects. These subtypes
were: (a) those with impairment in concept formation, in word
blending, verbal fluency and sentence memory, (b) those with impairment in sequencing, linguistic aspects and finger localization,
and (c) those with deficits in visual-spatial abilities, in verbal
comprehension and verbal coding. A similar study by Fisk and Rourke
(1979) with older learning disabled individuals found a high degree
of persistence across ages of the subtypes which were found to
account for 80 percent of the subjects. The Petrauskas and Rourke
(1979) and Fisk and Rourke (1979) findings are similar to those of
Mattis, French, and Rapin (1975). The latter group found three
syndromes which accounted for 90 percent of the variance in the 82
dyslexic children studied. The syndromes were: (a) language disorder, (b) dys-coordination in articulatory and graphomotor activities, and (c) visuo-constructional difficulties. Thus the above
subtype studies of dyslexics all add to our understanding of the
nature of the problem.

In a recent critical review and empirical study Satz and
Morris (1981) used cluster analysis to identify initially the
learning disabled prior to the search for the subgroups, which were
then validated against external criteria. Five distinctive clusters
or subtypes emerged: (a) those with global language difficulties,
(b) those with selective language difficulties such as verbal
fluency, (c) those with both language and perceptual handicaps,
(d) those selectively impaired on the nonlanguage perceptual tests,
and (e) the "unexpected learning disabled" with no impairment on the
neuropsychological tests. Satz and Morris are careful to emphasize
the preliminary nature of their findings. Their following concerns
apply equally well to other subtype studies: the need for wider
sampling of reading skills and of neuropsychological measures to
test these skills, the wider coverage in terms of age, sex and
ethnic background of the subjects and the strengthening of criterion
measures.

From the above review of current advances in subtyping of disabled readers or learning disabled children, it is clear the problem is a complex one and yet one worthy of continued study. Results
thus far have elucidated our understanding of these individuals.

ORTHOGRAPHY DIFFERENCES

From Gray's (1956) classic cross-national study of reading and
writing to Downing's (1973) comparative reading in thirteen
countries and from Gibson and Levin's (1975) to Downing and Leong's
(1982) psychology of reading to the empirical work on cross-
linguistic studies of reading and dyslexia (Kavanagh & Venezky,
1980) there is evidently a need to understand reading and reading
difficulties in different writing systems. Empirical findings will
explain similarities and differences in information processing
strategies during reading different orthographies and the levels or
stages at which these similarities and differences occur. It is
likely that with such disparate writing systems as English and
Chinese there are micro differences in processing English or Chinese
lexicon more than there are macro differences in processing prose
materials (see Leong, 1978). These processing strategies are also
reflected in hemispheric specialization. For example, Sasanuma
(1975) implicated superior right-hemisphere performance by Japanese
subjects in recognizing ideographic symbols. Tzeng and Hung (1980),
however, have argued from their studies of Chinese subjects reading
Chinese characters that laterality effects from visual half-field
experiments more likely reflect hemispheric differences in memory
storage location rather than cognitive processing in specialized
cortical areas. There is thus another dimension harking back to
neuropsychological studies, but with a different writing system.
Quite apart from the findings of hemispheric specialization, information processing paradigms with different types of writing systems
are beginning to yield explanations for processes of reading (see
Hung & Tzeng, 1981) and would likely be extended to aspects of
reading difficulties. As an important example, evidence is emerging
(see Kavanagh & Venezky, 1980) that phonological recoding is needed
at the working memory stage in early reading quite independently of
orthography. The representation of print, sound and meaning needs
to be further explored in both phonologically deep and shallow
languages.

ROLE OF LANGUAGE AWARENESS

To help disabled readers we need to develop in them an understanding of both the structure and function of language. This
relates to their language awareness which is variously referred to
as "metacognition," "reflective abilities," "general development of
consciousness" over and above knowing language as a formal system.
The term is well explained by Cazden (1974, p. 29):

> Metalinguistic awareness, the ability to make language
> form opaque and attend to them in and for themselves, is
> a special kind of language performance, one which makes

special cognitive demands, and seems to be less easily
and less universally acquired than the language performance of speaking and listening...

In their work on the psychology of reading Downing and Leong (1982,
Chapter 6) have discussed in detail the theoretical basis and
empirical studies relating language awareness to reading and reading
difficulties. They suggest that there are two main strands: the
need to understand the purposes of reading and the importance of
knowing the orthography (see also Downing, this volume). Both concepts have generated testable hypotheses and have been found to
underpin learning to read.

Working within the above framework and also drawing on the research program of Liberman and her associates (Liberman, Shankweiler,
Liberman, Fowler, & Fischer, 1977), my students and I have attempted
to provide some empirical evidence on the way in which language
awareness affects reading acquisition. In an earlier project, Leong
and Haines (1978) studied beginning readers' analysis of words and
sentences. We examined the performance of children in grades 1, 2,
and 3 in segmenting syllables into phonemes, words into syllables,
concepts of phonological representation and understanding of sentence structure. We confirmed the Liberman et al. (1977) finding
of difficulties in phoneme segmentation as compared with syllable
segmentation. Further, children across the three grades found
"high complexity" sentences (e.g., "The not very well painted toys
want painting during holidays") more difficult to imitate accurately
than "low complexity" sentences (e.g., "A little girl by the pond
was watching some ducks"). What is more, awareness of phonological
representation of sounds in words contributes significantly to
reading performance in a canonical correlation. The canonical correlation between the linear combination of the set of language
awareness and reading tasks of .777 is significant. The relative
contributions of the language tasks to reading are: phonological
awareness (relative weight of 6); sentences of high complexity
(weight of 5.4); syllable segmentation (weight of 3.6); phoneme
segmentation (weight of 1), while sentences of low complexity contribute -1.2 of the weight. These results are taken to mean that,
for these children, their awareness of words and sentences is at the
subsidiary level. Their acquisition of verbal skills is facilitated
if their understanding is brought to the focal level. The reflection on and manipulation of words and sentences, which can be taught
in the form of word games and play activities, will go some way
towards helping early readers and disabled readers.

In two further studies Leong and Sheh (1982, in preparation)
have used multivariate techniques to tease out the relationship between simultaneous-successive processing, language awareness (understanding of language rules, ambiguities and incongruities) and

reading (word recognition) in grades 2 and 4 children. On the basis
of factor scores from the simultaneous-successive components and
scores in the different language awareness tasks and a word recognition test, analyses have proceeded in two directions. In one direction, we have found as hypothesized that those children who are high
on both the simultaneous and successive dimensions also perform
better on language awareness tasks and on reading and that there is
a main effect for grade. The other direction is the more important
one. By using stepwise multiple regression analysis and path
analysis we have attempted to make more explicit the theoretical
framework as schematized in Figure 1. From Figure 1 it can be seen
that the dimensions of simultaneous and successive cognitive processing, as reflected in their factor scores, have very little
direct effect on reading as shown by the path coefficients of .0502
and .1150 respectively. Cognitive processing, however, affects
reading through its relationship with language awareness as shown
on the "indirect" path coefficients of .3039 and .3314 respectively.
It is clear that the language awareness tasks have a much greater
direct effect (path coefficient of .7351) on reading. By analyzing
for the direct and indirect effects of cognitive processing and
language awareness on reading, the Leong and Sheh studies thus carry
the earlier findings one step further.

In another study which is still in progress, I have followed
the high simultaneous-successive and the low simultaneous-successive
children, who were tested when they were in grades 2 and 4, into
their respective grades 3 and 5 in an attempt to further validate
the effects of language awareness on reading comprehension. The
differential performance of these children in the three types of
ambiguous sentences (lexical, surface and deep ambiguities) will

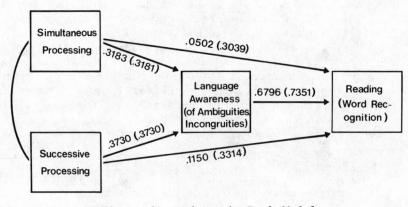

Figure 1.   Schematic Path Model

also be compared. Furthermore, a new group of grades 4 and 6 children have also been tested on their appreciation of ambiguities and
reading comprehension. Thus far, the initial results tend to col-
laborate and extend the Leong and Haines and Leong and Sheh studies.
Further work is planned to study the comprehension of ambiguities
by dyslexics compared with their controls to determine the effects
of this level of language awareness on reading and its breakdown.

Even though the above studies relate to unselected readers
there are clear research and teaching implications for disabled
readers. Language awareness is a mental activity which interacts
with other cognitive activities on which it depends and which it
can modify in turn. Studies of awareness usually refer to situations where the child will perform an action the results of which
are obvious to him or her and where the child is encouraged to
verbalize this action. This brings us back to the earlier discussion of learning to comprehend and teaching learning strategies.
From the remediation point of view, reading should be treated as
high-level self-motivating thinking activities. The role of the
teacher or the parent can at best be that of an external guide or a
facilitating agent in helping the child to develop a spontaneous
interest in reading. This interest will in turn provide its own
reinforcement. The need for a child-controlled rather than teacher-
dominated approach to reading is all the greater for disabled
readers. They need encouragement and guidance to adapt, correct,
judge, reflect on language and to know about language. For their
part, teachers need to teach language and reading not so much as a
rigid, formalized system but as an activity to be enjoyed, to be
manipulated and to be brought to some level of consciousness. The
emphasis is on both structure and function of language.

SUMMARY

Almost fifteen years ago Critchley (1968) suggested some seven-
teen areas worthy of further investigation in specific reading disability. These areas included: early and efficient diagnosis,
study of speech and language development, detailed case studies and
followups, studies of theory-based programs and remediation, research into brain-behavior mechanisms and information-processing in
disabled readers. These areas of investigation are still appropriate. As is evident from the present review and from other chapters
in this volume, progress has been made in many of the areas suggested by Critchley. And more. From a review of current work and
a projection into future ones, it seems reasonable to explicate the
concept of learning disabilities in the different academic areas.
In concentrating on reading difficulties we need to delineate the
relationship between micro analysis including automatic decoding and
macro analysis of text materials and to focus at greater depth on

reading comprehension and learning to comprehend strategies. This emphasis on micro and macro analyses should be within a neuropsychological framework and taking into account different writing systems and the central role of language awareness. The above relevant questions and promising areas of research are suggestive rather than exhaustive. They can never be, while there are still many Marys and Johnnys with reading or learning disabilities.

References

Anderson, R. C.   The notion of schema and the educational enterprise: General discussion of the conference. In R. C. Anderson, R. J. Spiro, and W. E. Montague (Eds.), Schooling and the acquisition of knowledge. Hillsdale, New Jersey: Erlbaum, 1977.

Bakker, D. J.   Perceptual asymmetries and reading proficiency. In M. Bortner (Ed.), Cognitive growth and development: Essays in honor of Herbert G. Birch. New York: Brunner-Mazel, 1979.

Baron, J., Treiman, R., Wilf, J. F., & Kellman, P.   Spelling and reading by rules. In U. Frith (Ed.), Cognitive processes in spelling. New York: Academic Press, 1980.

Bartlett, F. C.   Remembering. Cambridge: Cambridge University Press, 1932.

Benton, A. L., & Pearl, D. (Eds.), Dyslexia: An appraisal of current knowledge. New York: Oxford University Press, 1978.

Berger, N. S., & Perfetti, C. A.   Reading skill and memory for spoken and written discourse. Journal of Reading Behavior, 1977, 9, 7-16.

Bloomfield, L.   Language. New York: Holt, Rinehart & Winston, 1933.

Calfee, R. C.   Assessment of independent reading skills: Basic research and practical applications. In A. S. Reber and D. L. Scarborough (Eds.), Toward a psychology of reading. Hillsdale, New Jersey: Erlbaum, 1977.

Calfee, R. C.   Sources of dependency in cognitive processes. In D. Klahr (Ed.), Cognition and instruction. Hillsdale, New Jersey: Erlbaum, 1976.

Calfee, R. C.   Memory and cognitive skills in reading acquisition. In D. Duane and M. Rawson (Eds.), Reading, perception and language. Baltimore, Maryland: York Press, 1975.

Calfee, R. C., Chapman, R., & Venezky, R.   How a child needs to think to learn to read. In L. W. Gregg (Ed.), Cognition in learning and memory. New York: Wiley, 1972.

Calfee, R. C., & Drum, P. A.   Learning to read: Theory, research and practice. Curriculum Inquiry, 1978, 8, 183-249.

Calfee, R. C., & Spector, J. E.   Separable process in reading. In F. J. Pirozzolo and M. C. Wittrock (Eds.), Neuropsychological and cognitive processes in reading. New York: Academic Press, 1981.

Carr, T. H.   Building theories of reading ability: On the relation between individual differences in cognitive skills and reading comprehension. Cognition, 1981, 9, 73-114.

Cazden, C. B.   Metalinguistic awareness: One dimension of language experience. The Urban Review, 1974, 7, 28-39.

Chall, J., & Mirsky, A. F. (Eds.), Education and the brain. Chicago: University of Chicago Press, 1978.

Critchley, M.   Topics worthy of research. In A. H. Keeney and V. T. Keeney (Eds.), Dyslexia: Diagnosis and treatment of reading disorder. St. Louis: C. V. Mosby, 1968.

Cromer, W.   The difference model: A new explanation for some reading difficulties. Journal of Educational Psychology, 1970, 61, 471-483.

Cromer, W., & Wiener, M.   Idiosyncratic response patterns among good and poor readers. Journal of Consulting Psychology, 1966, 30, 1-10.

Cruickshank, W. M.   When winter comes, can spring ...? The Exceptional Child, 1978, 25, 3-25.

Dansereau, D.   The development of a learning strategies curriculum. In H. F. O'Neil (Ed.), Learning strategies. New York: Academic Press, 1978.

Das, J. P., Kirby, J., & Jarman, R. F.   Simultaneous and successive cognitive processes. New York: Academic Press, 1979.

Doehring, D. G.   Patterns of impairment in specific reading disability. Bloomington: Indiana University Press, 1968.

Doehring, D. G., & Hoshko, I. M.   Classification of reading problems by the Q-technique of factor analysis. Cortex, 1977, 13, 281-294.

Downing, J.   (Ed.), Comparative reading. New York: Macmillan, 1973.

Downing, J., & Leong, C. K.   Psychology of reading. New York: Macmillan, 1982.

Ellis, N. C., & Miles, T. R.   A lexical encoding deficiency I: Experimental evidence. In G. Th. Pavlidis and T. R. Miles (Eds.), Dyslexia research and its applications to education. New York: Wiley, 1981.

Farnham-Diggory, S.   Learning disabilities. Cambridge, Mass.: Harvard University Press, 1978.

Fillmore, C. J.   Some problems for grammar. In R. J. O'Brien (Ed.), Report of the Twenty-Second Round Table Meeting of linguistics and language studies. Washington, D.C.: Georgetown University Press, 1971.

Fillmore, C. J.   The case for case. In E. Bach and R. T. Harms (Eds.), Universals in linguistic theory. New York: Holt, Rinehart and Winston, 1968.

Fisk, J. L., & Rourke, B. P.   Identification of subtypes of learning-disabled children at three age levels: A neuropsychological, multivariate approach. Journal of Clinical Neuropsychology, 1979, 1, 289-310.

Frith, U.   Unexpected spelling problems. In U. Frith (Ed.),
    Cognitive processes in spelling. New York: Academic Press, 1980,
    (a).

Frith, U.   Cognitive processes in spelling. New York: Academic
    Press, 1980, (b).

Frith, U.   From print to meaning and from print to sound or how to
    read without knowing how to spell. Visible Language, 1978, 12,
    43-54.

Frith, U., & Frith, C.   Relationships between reading and spelling.
    In J. F. Kavanagh and R. L. Venezky (Eds.), Orthography, reading,
    and dyslexia. Baltimore: University Park Press, 1980.

Gibson, E. J., & Levin, H.   The psychology of reading. Cambridge,
    Mass.: MIT Press, 1975.

Goldman, S. R., Hogaboam, T. W., Bell, L. C., & Perfetti, C. A.
    Short-term retention of discourse during reading. Journal of
    Educational Psychology, 1980, 5, 647-655.

Gray, W. S.   The teaching of reading and writing. Paris: UNESCO,
    1956.

Grimes, J.   The thread of discourse. The Hague: Mouton, 1975.

Haines, L. P.   Visual and phonological coding in word processing by
    grades 4, 6, and 8 readers. Unpublished Doctoral Dissertation,
    University of Saskatchewan, 1978.

Halliday, M. A. K., & Hasan, R.   Cohesion in English. London:
    Longmans, 1976.

Hogaboam, T. W., & Perfetti, C. A.   Reading skill and the role of
    verbal experience in decoding. Journal of Educational
    Psychology, 1978, 70, 717-729.

Hung, D. L., & Tzeng, O. J. L.   Orthographic variations and visual
    information processing. Psychological Bulletin, 1981, 90, 377-
    414.

Kant, E.   (Critique of pure reason.) (K. Smith, trans.) London:
    Macmillan, 1963. (Originally published, 1781.)

Kavanagh, J. F., & Venezky, R. L.   (Eds.), Orthography, reading,
    and dyslexia. Baltimore: University Park Press, 1980.

Kintsch, W.   The representation of meaning in memory, New York:
    Wiley, 1974.

Kirk, S. A.   Behavioral diagnosis and remediation of learning
    disabilities. Proceedings of conference on exploration into the
    problems of the perceptually handicapped child. Chicago:
    Perceptually Handicapped Children, Inc., 1963.

Knights, R. M., & Bakker, D. J.   (Eds.), The neuropsychology of
    learning disorders: Theoretical approaches. Baltimore:
    University Park Press, 1976.

LaBerge, D., & Samuels, S. J.   Toward a theory of automatic infor-
    mation processing in reading. Cognitive Psychology, 1974, 6,
    293-323.

Legein, Ch. P., & Bouma, H.   Visual recognition experiments in
    dyslexia. In G. Th. Pavlidis and T. R. Miles (Eds.), Dyslexia
    research and its applications to education. New York: Wiley,
    1981.

Leong, C. K.   Cognitive strategies in relation to reading
    disability. In M. P. Friedman, J. P. Das, and N. O'Connor (Eds.),
    Intelligence and learning. New York: Plenum, 1981.
Leong, C. K.   Laterality and reading proficiency in children.
    Reading Research Quarterly, 1980, 15, 185-202. (a)
Leong, C. K.   Cognitive patterns of 'retarded' and below-average
    readers. Contemporary Educational Psychology, 1980, 5, 101-117.
    (b)
Leong, C. K.   Learning to read in English and Chinese: Some
    psycholinguistic and cognitive considerations. In D. Feitelson
    (Ed.), Cross-cultural perspectives on reading and reading
    research. Newark, Delaware: IRA, 1978.
Leong, C. K.   Spatial temporal information processing in children
    with specific reading disability. Reading Research Quarterly,
    1976-1977, 12, 204-215.
Leong, C. K.   Lateralization in severely disabled readers in
    relation to functional cerebral development and synthesis of
    information. In R. M. Knights and D. J. Bakker (Eds.), The
    neuropsychology of learning disorders: Theoretical approaches.
    Baltimore: University Park Press, 1976.
Leong, C. K., & Haines, C. F.   Beginning readers' awareness of
    words and sentences. Journal of Reading Behavior, 1978, 10,
    393-407.
Leong, C. K., & Sheh, S.   Knowing about language--some evidence
    from readers. Annals of Dyslexia (formerly) Bulletin of the
    Orton Society), 1982, 32 (in press).
Leong, C. K., & Sheh, S.   Cognitive processing, language awareness
    and reading in grade 2 and grade 4 children (manuscript in
    preparation).
Liberman, I., Shankweiler, D., Liberman, A. M., Fowler, C., &
    Fischer, F. W.   Phonetic segmentation and recoding in the
    beginning reader. In A. S. Reber and D. L. Scarborough (Eds.),
    Toward a psychology of reading. Hillsdale, New Jersey: Erlbaum,
    1977.
Luria, A. R.   Human brain and psychological processes. New York:
    Harper and Row, 1966.
Markman, E. M.   Comprehension monitoring. In W. P. Dickson (Ed.),
    Children's oral communication skills. New York: Academic Press,
    1981.
Mason, M.   Reading ability and the encoding of item and location
    information. Journal of Experimental Psychology: Human
    Perception and Performance, 1980, 6, 89-98.
Mason, M.   From print to sound in mature readers as a function of
    reading ability and two forms of orthographic regularity.
    Memory and Cognition, 1978, 6, 568-581.
Mason, M., Pilkington, C., & Brandau, R.   From print to sound:
    Reading ability and order information. Journal of Experimental
    Psychology: Human Perception and Performance, 1981, 7, 580-
    591.

Mattingly, I. G.   Reading, the linguistic process, and linguistic awareness.   In J. F. Kavanagh and I. G. Mattingly (Eds.), Language by ear and by eye.   Cambridge, Mass.: MIT Press, 1972.

Mattis, S.   Dyslexia syndromes: A working hypothesis that works. In A. L. Benton and D. Pearl (Eds.), Dyslexia: An appraisal of current knowledge.   New York: Oxford University Press, 1978.

Mattis, S., French, J. H., & Rapin, I.   Dyslexia in children and young adults: Three independent neuropsychological syndromes. Developmental Medicine and Child Neurology,   1975, 17, 150-163.

Maxwell, A. E.   Factor analysis: Thomson's sampling theory re-called.   British Journal of Mathematical and Statistical Psychology,   1972, 25, 1-21. (a)

Maxwell, A. E.   The WPPSI: A marked discrepancy in the correlation of the subtests for good and poor readers.   British Journal of Mathematical and Statistical Psychology,   1972, 25, 283-291. (b)

Miller, G. A.   Toward a third metaphor for psycholinguistics.   In W. B. Weimer and D. S. Palermo (Eds.), Cognition and the symbolic processes.   New York: Wiley, 1974.

Minsky, M.   A framework for representing knowledge.   In P. H. Winston (Ed.), The psychology of computer vision.   New York: McGraw-Hill, 1975.

Myklebust, H. R.   Progress in learning disabilities, Vol. IV. New York: Grune & Stratton, 1978.

Myklebust, H. R., Bannochie, M., & Killen, J.   Learning disabilities and cognitive processes.   In H. R. Myklebust (Ed.), Progress in learning disabilities, Vol. II.   New York: Grune & Stratton, 1971.

Naidoo, S.   Specific dyslexia.   London: Pitman, 1972.

Nelson, H. E.   Analysis of spelling errors in normal and dyslexic children.   In U. Frith (Ed.), Cognitive processes in spelling. New York: Academic Press, 1980.

Nelson, H. E., & Warrington, E. K.   Developmental spelling retardation and its relation to other cognitive abilities. British Journal of Psychology,   1974, 65, 265-274.

Oaken, R., Wiener, M., & Cromer, W.   Identification, organization, and reading comprehension for good and poor readers.   Journal of Educational Psychology,   1971, 62, 71-78.

O'Neil, H. F.   (Ed.), Learning strategies.   New York: Academic Press.

Pascarella, E. T., & Pflaum, S. W.   The interaction of children's attributions and level of control over error correction in reading instruction.   Journal of Educational Psychology,   1981, 73, 533-540.

Patterson, K. E.   Neuropsychological approaches to the study of reading.   British Journal of Psychology,   1981, 72, 151-174.

Pearson, P. D., & Spiro, R. J.   Toward a theory of reading compre-hension instruction.   Topics in Language Disorders,   1980, 1, 71-88.

Perfetti, C. A.   Language comprehension and fast decoding: Some psycholinguistic prerequisites for skilled reading comprehension. In J. T. Guthrie (Ed.), Cognition, curriculum, and comprehension. Newark, Delaware: International Reading Association, 1977.

Perfetti, C. A., Finger, E., & Hogaboam, T.   Sources of vocalization latency differences between skilled and less skilled young readers. Journal of Educational Psychology, 1978, 70, 730-739.

Perfetti, C. A., & Goldman, S. R.   Discourse memory and reading comprehension skill. Journal of Verbal Learning and Verbal Behavior, 1976, 14, 33-42.

Perfetti, C. A., & Hogaboam, T.   Relationship between single word decoding and reading comprehension skill. Journal of Educational Psychology, 1975, 67, 461-469.

Perfetti, C. A., & Lesgold, A. M.   Discourse comprehension and sources of individual differences. In P. Carpenter and M. Just (Eds.), Cognitive processes in comprehension. Hillsdale, New Jersey: Erlbaum, 1977.

Petrauskas, R. J., & Rourke, B. P.   Identification of subtypes of retarded readers: A neuropsychological, multivariate approach. Journal of Clinical Neuropsychology, 1979, 1, 17-37.

Pflaum, S. W., & Bryan, T. H.   Oral reading research and learning disabled children. Topics in Learning and Learning Disabilities, 1982, 1, 33-42.

Pflaum, S. W., & Pascarella, E. T.   Interactive effects of prior reading achievement and training in context on the reading of learning disabled children. Reading Research Quarterly, 1980, 16, 138-158.

Piaget, J.   The language and thought of the child. London: Routledge and Kegan Paul, 1926.

Pike, K. L.   Language in relation to a united theory of the structure of human behavior. The Hague: Mouton, 1967.

Resnick, L. B., & Glaser, R.   Problem solving and intelligence. In L. B. Resnick (Ed.), The nature of intelligence. New York: Wiley, 1976.

Rourke, B. P.   Reading, spelling, arithmetic disabilities: A neuropsychologic perspective. In H. R. Myklebust (Ed.), Progress in learning disabilities, Vol. IV. New York: Grune & Stratton, 1978.

Rourke, B. P., & Finlayson, M. A. J.   Neuropsychological significance of variations in patterns of academic performance: Verbal and visual-spatial abilities. Journal of Abnormal Child Psychology, 1978, 6, 121-133.

Rourke, B. P., & Gates, R. D.   Neuropsychological research and school psychology. In G. W. Hynd and J. E. Obrzut (Eds.), Neuropsychological assessment and the school-age child. New York: Grune & Stratton, 1981.

Rourke, B. P., & Strang, J. D.   Neuropsychological significance of variations in patterns of academic performance: Motor, psychomotor, and tactile-perceptual abilities. Journal of Pediatric Psychology, 1978, 3, 62-66.

Rumelhart, D. E.   Understanding and summarizing brief stories.   In
    D. LaBerge and S. J. Samuels (Eds.), Basic processes in reading:
    Perception and comprehension. Hillsdale, New Jersey: Erlbaum,
    1977. (a)
Rumelhart, D. E.   Toward an interactive model of reading.   In S.
    Dornic (Ed.), Attention and performance VI. Hillsdale, New
    Jersey: Erlbaum, 1977. (b)
Sasanuma, S.   Kana and Janji processing in Japanese aphasics.
    Brain and Language, 1975, 2, 369–383.
Satz, P., & Morris, R.   Learning disability subtypes: A review.
    In F. J. Pirozzolo and M. C. Wittrock (Eds.), Neuropsychological
    and cognitive processes in reading. New York: Academic Press,
    1981.
Schank, R., & Abelson, R.   Scripts, plans, goals and understanding.
    Hillsdale, New Jersey: Erlbaum, 1977.
Seymour, P. H. K., & Porpodas, C. D.   Lexical and non-lexical
    processing of spelling in dyslexia. In U. Frith (Ed.),
    Cognitive processes in spelling. New York: Academic Press, 1980.
Spiro, R. J., Bruce, B. C., & Brewer, W. F.   Theoretical issues in
    reading comprehension. Hillsdale, New Jersey: Erlbaum, 1980.
Spreen, O.   Post-conference summary. In R. M. Knights and D. J.
    Bakker (Eds.), The neuropsychology of learning disorders:
    Theoretical approaches. Baltimore: University Park Press, 1976.
Stanovich, K. E.   Toward an interactive-compensatory model of
    individual differences in the development of reading fluency.
    Reading Research Quarterly, 1980, 16, 32–71.
Steiner, R., Wiener, M., & Cromer, W.   Comprehension training and
    identification for poor readers and good readers. Journal of
    Educational Psychology, 1971, 62, 506–513.
Thorndike, P. W.   Cognitive structures in comprehension and memory
    of narrative discourse. Cognitive Psychology, 1977, 9, 77–110.
Tzeng, O., & Hung, D.   Reading in a nonalphabetic writing system:
    Some experimental studies. In J. F. Kavanagh and R. L. Venezky
    (Eds.), Orthography, reading, and dyslexia. Baltimore:
    University Park Press, 1980.
U.S. Office of Education. Assistance to states for education of
    handicapped children, notice of proposed rulemaking. Federal
    Registrar, 1976, 41 (No. 230), 52404–52407.
Vellutino, F. R.   Dyslexia: Theory and research. Cambridge,
    Mass.: MIT Press, 1979.
Vernon, M. D.   Variability in reading retardation. British Journal
    of Psychology, 1979, 70, 7–16.
Woods, S. S., Resnick, L. B., & Groen, G. J.   An experimental test
    of five process models for subtraction. Journal of Educational
    Psychology, 1975, 67, 17–21.
Young, A. W., & Ellis, A. W.   Asymmetry of cerebral hemispheric
    function in normal and poor readers. Psychological Bulletin,
    1981, 89, 183–190.

RESEARCH IN LEARNING DISABILITIES: A VIEW OF STATUS AND NEED[1]

Barbara K. Keogh

Department of Education
University of California, Los Angeles

Research investigators and clinicians working with children
with learning disabilities (LD) have identified a broad range of
intriguing and difficult problems. Pursuit of these problems has
led to an extensive and ever increasing volume of literature and has
spawned a variety of intervention approaches. It might be argued
that the most difficult problem facing the researcher of learning
disabilities is to decide what should be studied and who should be
included in the study sample. That is, what constitutes the field
and what constitutes the condition?

Consider. As researchers we seek to provide systematic evidence on a condition in which incidence figures vary from 2 to 30%
of the regular population. A variety of professional disciplines
are represented: Psychology, Education, Neurology, Psychiatry,
Opthalmology, Optometry, and increasingly (sometimes to our dismay),
Law. Identification methods vary widely. In a recent review of 408
studies Keogh, Major, Omori, Gandara, and Reid (1980) found that
over 1400 diagnostic techniques had been used to select or describe
learning disabled children. These included 40 different IQ tests
and 79 different achievement tests. To illustrate the diversity of
techniques, included in the range of tests used specifically for
selection purposes were: the Navy Ataxia Battery, the WISC-R, the
Florentino-Reflex Test, the Wagner Oral Inventory for Reversals, and
the Bender Gestalt. Children classified as learning disabled in the

---
[1] I wish to thank Susan Major-Kingsley, Helen Patricia Reid, and Lisa
Omori-Gordon for their help in gathering material for this paper and
for their continuing consultation and collaboration.

27

studies reviewed presented a variety of symptoms, as for example:
underachievement, clumsiness, dysdiadokinesis, sleep disorders,
depression, obesity, and hyperactivity. Classification terms
included: brain injured, pseudo-backward, sleep apneic, slow
learner, aphasic, and neurologically impaired. Research subjects
were selected from a number of sources which differed dramatically,
some researchers studying teacher identified children, others selecting hospital-inpatients.

To compound the problem further, therapeutic or intervention
approaches vary widely in philosophy and in practice. Even cursory
review of the literature suggests that there are few comprehensive
studies which deal directly with questions of program impact or in
which evaluation data are systematically collected. Possible interactions of program and child characteristics (Aptitude Treatment
Interactions) cannot be tested because the characteristics of both
subjects and programs are uncertain.

In short, subjects, diagnostic methods, and intervention all
reflect the disciplinary training, perhaps even the vested interests, of those who study or work with learning disabled children.
While there are certainly some common areas of interest and concern,
it appears that each investigator takes a "little slice" from his or
her particular perspective. The different views may all be important and in many instances are likely related; yet each study involves children with somewhat different attributes and problems, and
each investigator brings a somewhat different definition of learning
disabilities to the research task. The result of this variability
is a limited data base and a lack of generalizability of findings
across studies. No single, coherent theoretical position dominates
the field. Despite research and practice for almost 20 years, we
still lack a consistent and agreed upon frame of reference within
which to conduct systematic and comprehensive research.

UCLA MARKER VARIABLE PROJECT

Subject variability and definitional inconsistencies were
studied directly in the UCLA Marker Variable Project already referred to (Keogh, Major, Reid, Gandara, & Omori, 1978; Keogh et al.,
1980; Keogh, Major-Kingsley, Omori-Gordon, & Reid, 1981). As data
from this work provide the basis for much of the discussion in the
present chapter, the project will be described briefly. This 3 year
project, funded by the Bureau of Education for the Handicapped, U.S.
Office of Education, was aimed at developing a system of subject
descriptors which could be used by researchers when reporting work
with learning disabled subjects. Use of a marker system provides a
basis for determining comparability of subjects within and across
studies and allows the alignment of findings from different studies.
Use of a marker system might even lead to aggregating or combining
data from different geographic areas.

The UCLA project was carried out in three stages. The first involved a comprehensive and detailed review of subject descriptors found in the empirical literature in learning disabilities from 1970 through 1977. This literature search, involving both hand and computer techniques, yielded over 4600 citations, approximately 1360 of which were subject or data based. On the basis of this review and after consideration of major conceptual positions, commonly used subject descriptors, or "markers," were tentatively identified and defined operationally. In general, markers were defined as sample or subject variables which could be used as "... descriptive 'bench marks' or common reference points" (Keogh et al., 1981). Markers "... reflect the constructs which define and characterize a partic- ular field, and thus provide both operational comparability and conceptual organization" (Keogh et al., 1978, p. 7). Following critical discussion with a series of active researchers in the learning disabilities field, during Phase Two the markers were refined and placed into three major groups: Descriptive Markers, Substantive Markers, and Topical Markers; a fourth group, Background Information Markers, were identified in Phase Three. A list of markers may be found in Table 1. A Marker Guide was developed for practicality and the markers and Guide were field tested by investi- gators conducting research on learning disabilities. Final revi- sions were made on the basis of the field test data. The result is a working Guide feasible for use in learning disabilities research.

Table 1. Markers by Category

Descriptive Markers

Number of subjects by sex
Chronological age
Grade level
Locale
Race/ethnicity
Source of subjects
Socioeconomic status
Language
Educational history
Educational placement
Physical and health status

Substantive Markers

Intellectual ability
Reading achievement
Arithmetic achievement
Behavioral and emotional
  adjustment

Topical Markers

Activity level
Attention
Auditory perception
Fine motor coordination
Gross motor coordination
Memory
Oral language
Visual perception

Background Markers

Year of study
Geographical location
Exclusionary criteria
Control/comparison group

While the Marker Guide and the frame of reference provided by a marker system may be useful for researchers and program specialists in the field, of particular relevance to this paper are the major research problems and limitations which were identified in the literature search and field test. These research problems may provide direction for future work in the field.

LIMITATIONS IN LD RESEARCH

Based on our comprehensive review of the empirical literature and the field test data collected in the Marker Project, several generalizations may be drawn about learning disabilities research to date. First, we face a major and continuing problem of sample heterogeneity. As a consequence, it is almost impossible to draw solid generalizations from any single study. Further, sample heterogeneity quite clearly limits consideration of the comparability of findings across studies. Second, current research is almost exclusively child-focused. A broad range of particular characteristics has been studied, but for the most part potentially important extra-child variables such as contextual or situational influences have been ignored. Third, and closely related, we have ignored the obvious: It is, indeed, a multivariate world, yet most studies are focused on single variables.

Each of these problems has a somewhat different emphasis, yet it may be that they stem from a common problem; that is, the lack of an adequate theory from which to derive an agreed upon organizational system or taxonomy within the field. The categories of learning disabilities and the sub-areas of problems reflect various professional disciplines rather than logical or empirical structure. It is difficult to tell which are broad classification terms and which are specific subunits. It is therefore extremely difficult to determine what is subsumed by what. Criteria for class membership are ill defined, and therapeutic implications from diagnosis are inconsistent. In short, our inability to form a coherent conceptual system has limited the field both in terms of its research and its applied aspects. While obviously related, the three major limitations identified in the Marker Project literature search (sample heterogeneity, exclusive child-focus, and univariate strategies) produce somewhat different problems. Each deserves brief discussion.

Sample Heterogeneity

Two kinds of sample heterogeneity have affected research in learning disabilities. The first relates to subject differences across studies; the second focuses upon subject differences within studies. Recalling the many different professional disciplines involved in the study and treatment of learning disabilities, it is

not surprising that individual researchers use somewhat different
criteria and techniques for selecting learning disabled subjects.
Clearly, there are a number of possible bases for classification
and identification.

Investigators who choose a neurological perspective might select
on evidence believed to implicate causes of the condition, e.g.,
prenatal history, perinatal conditions, history of injury or illness,
etc. This approach has been referred to as the "medical model" in
which diagnosis of cause is likely to determine treatment, and in
which learning disabilities are presumed to have a physical or neurological basis. In contrast, another classification and selection
system might be based on educational competence, without regard for
cause or etiology; in such an approach the major definitional and
selection parameters would be achievement and ability indicators,
and an ability-achievement index might be the basis for inclusion in
the study sample. Still another basis of classification might be
evidence of disorders in psychological processes; that is, individual subjects are identified according to "auditory processing
deficits" or "visual perceptual disorders." Yet another approach
might focus on behavior without regard for etiological, physical,
educational, or psychologically inferred parameters. The point to
be made is that whatever the specifics, identification criteria lead
to selection of partially overlapping but somewhat different groups.
The goal of the individual researcher is to reduce sample variability and increase the homogeneity of his/her sample. Yet,
because the identification and selection processes reflect different
theoretical and classification perspectives, the subjects in one
study or program may be dramatically different from those in
another. It should not surprise us, then, that similar experimental
or intervention procedures often lead to different findings, and
that a broad array of subjects carry the learning disabled label.

A second aspect of sample heterogeneity has to do with differences in subjects within samples. In many research efforts subjects
are identified and considered eligible for classification as learning disabled on the basis of a limited set of characteristics, e.g.,
all children selected had IQs between 80-95 and were two years below
grade level in reading as determined by a standardized reading test.
Or, all selected children evidenced neurological soft signs and were
hyperactive in the classroom as described by the teacher. Or, all
children were in regular classes and had visual perceptual problems
as demonstrated on the Bender Gestalt Test. Despite similarity on
particular selection criteria, the assumption of within sample homogeneity is suspect for several reasons. First, subject similarity
on one variable or characteristic does not guarantee similarity on
others. Second, the attributes, symptoms, or characteristics which
determine selection may change with age and development. Third,
classification and selection is a function of the particular measurement technique or test used, and thus may be flawed because of
psychometric limitations of the instrument.

Problems of assessment and measurement in learning disabilities
research are clearly illustrated by the UCLA Marker Variable data in
which we found over 440 tests used in subject selection and description. Over 40 measures of IQ alone and over 70 achievement measures
were found in our review of 408 studies. More importantly, perhaps,
only three IQ tests were used more than eight times by the large
sample of participants in this Project. Review of the tests against
criteria for psychometric adequacy (see Salvia & Ysseldyke, 1978)
suggests that few met even minimal measurement requirements. The
limits of measurement, then, are very real. The psychometric properties of most instruments used in learning disabilities work are
not known and their age appropriateness is questionable in a good
deal of the published research. It is important to note, too, that
the factor structure of psychological abilities may vary by age,
thus raising issues of interpretability of test scores in research
or clinical applications. Given the vagaries of identification and
selection techniques, both theoretically and practically, it is not
surprising that sample heterogeneity is the norm rather than the
exception. Still another aspect of the problem deserves brief
discussion.

Extra-child influences

In addition to test limitations, sample inconsistency is compounded by a second problem identified in the UCLA Marker Project:
learning disabilities have been viewed as an in-child condition,
when, more realistically, the condition is a function of both child
and extra-child variables. A child's development and competencies
are influenced by many environmental and experiential variables as
well as by diverse in-child characteristics. Lack of concern or
recognition of the social, educational, and situational influences
on children's development and performance may lead to errors in selection and classification. In particular, it may result in exclusion of children with learning disabilities.

To illustrate. In many formal definitions of learning disabilities we have chosen to separate lack of achievement related to
"social or cultural disadvantage" from presumed in-child conditions,
as if these two sources of variability were mutually exclusive.
Yet, it is abundantly clear that a child's abilities and problems
are influenced and affected by the nature of his experiences as well
as by his biological potential or neurological condition. "Matu-
rational lag" or "minimal brain dysfunction" may have different
expressions and different long term consequences as a function of
the family and social group in which the child is raised. Compounding effects of parents' attitudes and behaviors, the achievement climate of the home, and the opportunities for help and intervention all enhance individual differences in children's development. The nature of the instructional program of the school and

classroom environments influence educational and personal competence.
All may be positive or negative contributors to the learning disabilities condition. While few would disagree with that interpretation,
it is interesting to note that there has been little systematic research directed at extra-child influences. With few exceptions
(Adelman, 1971; Werner, Bierman, & French, 1971; Werner, 1980) there
has been a remarkable lack of concern for these powerful situational
influences in the study of learning disabilities. In our efforts to
simplify the research and intervention tasks we may have ignored the
complexity of the condition by focusing exclusively on the characteristics of children, and not on their home and school environments.

Although not addressed in this chapter, the possible influence
of the content and structure of educational tasks on the performance
of learning disabled children deserves consideration. The organization of curricular content, and the order and sequence of presentation, may have important consequences for children's accomplishments,
especially when there are problems in learning. The nature of the
interactions of child characteristics and task characteristics are
unclear, but provide potentially powerful approaches to the understanding of learning disabled children and to appropriate interventions. The nature of the learning task represents still another
source of variability which must be considered when studying learning disabilities.

A multivariate Condition

It might be argued that lack of concern for extra-child variables relates to a tendency to ignore the multivariate nature of the
learning disabilities condition. Yet, as in most human conditions,
development is "overdetermined" and awareness of the contribution of
many sources of influence is required. Not surprisingly, most researchers find it easier to deal with a limited number of study
variables than to try to integrate a variety of influences. In some
cases the nature of the variables may be uncertain and their interactions unknown.

As with the extra-child influences already noted, the problem
is in part related to measurement limitations. There are few operational measures of many important child attributes or of settings.
While a number of investigators suggest that learning disabled
children have serious problems in motivation, affect, and other
psychodynamic processes, the measurement problems in the motivational and affective areas are especially real. Lack of adequate
measurement of self-concept, for example, severely restricts
determination of the comparability of subjects within or across
samples on this particular dimension. Further, the compounding,
interactive effect of a number of variables is poorly understood,
and the relative weights of given influences are unknown.

Acceptance of Complexity

Taken as a whole, research in learning disabilities to date has
been limited by problems of sample heterogeneity, by focus on child
characteristics, and by lack of concern for the multivariate nature
of the condition. There is still no agreed upon organization of the
field which yields a workable taxonomy. As a consequence the research tends to be fragmented and few major generalizations can be
drawn with confidence. As noted earlier in this chapter, these
problems likely stem, in part at least, from lack of powerful
theories or models with which to guide our research. Given the
applied nature of the field it seems likely that the conceptual
models useful in directing research will necessarily be drawn from
other disciplines. A number of approaches hold promise, but the
extensive work in developmental psychology may be closest to the
needs of researchers of learning disabilities. A particularly
attractive model is that proposed by Sameroff (1975).

DEVELOPMENTAL MODELS

Sameroff (1975) has described three models which purport to
explain the process of development; by inference these models may be
useful in understanding problems in development. The models may be
especially helpful in analyzing approaches to the study and treatment of learning disabilities.

The first, a main effect model, is one in which either constitutional or environmental factors are recognized as influences on
development; these influences are, however, viewed as independent of
each other. To illustrate, a child might be learning disabled
because he has neurological impairment; because he came from a particularly disturbed home, or because he had a poor teacher. Within
this model, a child might be neurologically impaired and also be
emotionally disturbed, but these influences are viewed as parallel,
at most, additive, not interactive.

A second model described by Sameroff has been called the interactional model. In this view, development is a function of contributions of both constitutional and environmental factors which
affect each other; that is, the child with neurological impairment
might be aided or harmed by the nature of the home environment or
by the strength of the teaching program. Both child and other variables, therefore, must be considered in understanding a learning
disabled condition.

While the interactional model is clearly a step beyond the main
effect model, Sameroff argues that, in fact, development is transactional. That is, that child and caretaker change each other; thus,

it is not possible to use a simple main effect model or even an interactive model to understand development. Said simply, in an interactive model the various sources of influence affect each other; in the transactional model they change each other. Following this logic, it could be argued that from the perspective of a transactional model the learning disabled child might indeed be neurologically impaired but that his condition in turn changes his environment, the modified environment then acting upon the child to produce further change, e.g., to exacerbate the neurological condition or to modify it in other ways. The transactional approach is well illustrated in the thorough review of followup studies of infant risk conditions by Sameroff and Chandler (1975). Defining infant risk as anoxia, prematurity, delivery complications, and social conditions, the authors provide a compelling argument to suggest that the most significant factor in outcome for risk infants is not just the biological risk condition (i.e., anoxia or prematurity), but the nature of the transactions with the social milieu in which the infant lives. The transactional process is well illustrated, too, in Kearsley's (1979) discussion of iatrogenic retardation, a condition viewed as a form of "learned incompetence." Kearsley makes a strong case for the mutually reinforcing yet transforming effects of child and family, so that the child's problems lead to changes in the behaviors of parents which, in turn, lead to further limitations in the child's development.

Applied to learning disabilities the transactional model carries a number of implications which include aspects of diagnosis and identification as well as of treatment. First, it implies that learning disabilities may not be due exclusively to stable in-child characteristics, but rather that learning disabilities are an expression of the transaction which has occurred between the child and his environment. Second, the transactional model suggests that we must expect changes in condition over time, as the child, family, or school environment mutually transform each other. Finally, it suggests that diagnostic and selection procedures must be broadened to include a variety of extra-child as well as in-child conditions.

RESEARCH NEEDS

Based on this kind of analysis what, then are the research needs to be addressed in the next several years? These needs particularly relate to sampling, timing, and methods. They are well illustrated by consideration of age, sex, and cultural differences in the expression of learning disabilities.

Age Differences

Despite the volume of published studies describing characteristics of learning disabled children, there are real questions about the expression of learning disabilities in different age groups and over time. As documented in the UCLA Marker Project (Keogh et al., 1980, 1981) the bulk of the data based information is on six- to twelve-year-old children. Examination of the research literature reveals some emerging trends, however. There are increasing numbers of studies directed at adolescents and young adults and some concern for preschool children with actual or potential learning problems. The latter direction has emerged from the continuing interest in early screening and identification programs. Although rather cynical, we might speculate that the concern for adolescents and young adults has come about because of lack of success in "curing" learning disabilities in younger age groups. In any case, the recent years have seen a broadening of research efforts to include samples selected from an age range preschool through young adult-hood. Unfortunately, a number of basic questions related to possible age effects remain unanswered.

The bulk of studies to date have utilized cross sectional approaches, describing learning disabilities within different age groups, and have provided important information. Yet it seems likely that real insight into the "natural course" of development of the condition, including the impact of interventions, will require longitudinal and followup designs. These approaches are demanding in time and are expensive. They also have serious problems related to sample attrition and to analytic techniques. As a consequence, there are only a limited number of comprehensive longitudinal studies of learning disabled children in the published literature. Importantly, these studies often yield contradictory evidence and lead to different conclusions.

As documented in a recent review by Major-Kingsley (1981), some investigators suggest that learning disabilities are "outgrown" or overcome by the time most children reach adulthood (Preston & Yarrington, 1967; Rawson, 1978; Robinson & Smith, 1962). Others argue that a considerable number of learning disabled children continue to have problems throughout their school years and into their adult lives (Balow & Blumquist, 1965; Gottesman, 1979). Pro-ponents of the latter view suggest that even though the specific expression of the problems (i.e., the symptoms) may change, problems in development remain.

Evidence for continuing problems in adjustment is reported in a number of studies where learning disabilities were associated with hyperactivity. Huessy, Metoyer, and Townsend (1974), as an example, conducted an eight to ten year followup of children who had been diagnosed as hyperkinetic and placed on medication. The authors concluded that hyperkinetic children remain at risk for academic, emotional, and social problems in later life. Subjects in their study did not "outgrow" their problems. There was a higher than

expected rate of school drop-out, institutionalization for delin-
quent behavior, and need for psychiatric treatment. These findings
are essentially consistent with the rather gloomy followup report
of Minde, Weiss, and Mendelson (1972), but are different from more
positive findings of Huessy and Gendron (1970), Hechtman, Weiss,
Finklestein, Werner, and Benn (1976). Most of the investigators
reporting positive long term outcomes do note, however, that many
of their hyperactive subjects continued to have some indications of
impulsivity, restlessness and/or poor socialization skills
(Ackerman, Dykman, & Peters, 1977a, b). Nevertheless, as described
in detail by Hechtman et al. (1976), hyperactives in their study
were similar to comparison subjects on a number of dimensions
(employment status, job satisfaction, use of drugs, psychiatric
referral rate, and antisocial behavior); educational attainment was
the major area of difference.

What might account for the differences in findings from longitudinal studies? It seems reasonable that some of the differences
relate to research design and methodological considerations. Several points deserve brief discussion. First, it is likely that there
are real differences in samples or subjects selected for longitudinal study (see Keogh et al., 1981, for detailed discussion). Some
longitudinal investigators draw their subjects from clinics or
hospital registers. Others choose teacher referred children, pupils
identified for special education services, or the lower 10-25th per-
centiles of the distribution of children on reading and arithmetic
tests. Some investigators limit their samples to boys; others
select both boys and girls. A number of researchers select children
early on in their school years, some going back as far as preschool;
others enter children into the study in the early adolescent period.
Stringent criteria relating to socio-economic status or ethnicity
are applied by some investigators, while others include all available subjects without regard to those selection criteria. In short,
major and serious sampling differences have probably influenced the
nature of the findings; thus, generalizations about the long term
consequences of learning disabilities must be drawn with real
caution. Given the heterogeneity of the longitudinal samples described in the published literature, it is not surprising that
findings and conclusions are discrepant.

Other influences on longitudinal or followup findings relate
to the nature of the outcome indices selected as well as to time to
followup. Some investigators view success as freedom from involvement with the judicial system; others look to adequate personal-
social adjustment, to vocational success, or to completion of school
as evidence of success. The variety of outcome measures is broad
and there is little consistency across studies or even across
measures. Closely related, time to followup varies significantly.
Some investigators consider 2 years as longitudinal; others have

followed children over their full school careers; still others study children from the elementary school years through young adulthood. As there is some instability of adjustment and development for any given child across age periods, it is not surprising that differences in time to followup yield somewhat different results.

Still another source of variability relates to a general lack of concern about the nature of the child's experiences between iden- tification and followup. Some children receive extensive inter- vention and specialized services including psychotherapy, education- al tutoring, and the like. Other children receive relatively little specific help, and where interventions were attempted, they were short term and unsystematic. Common sense as well as research evi- dence argue that the nature of intervention or intervening experi- ences are potent influences on children's development and thus deserve consideration if we are to interpret the findings of longi- tudinal or followup work. The point is especially important within the context of a transactional model as proposed by Sameroff (1975).

Finally, the nature of the design and statistical techniques applied to longitudinal data are complex. The relatively simple statistical demonstration of correlation between variables over time is not adequate, and refined and sophisticated analytic strategies are needed if the complex interactions of child and situational influences are to be delineated accurately. In this regard the life span work of Baltes and his associates (Baltes, Reese, & Nesselroade, 1977) deserve serious consideration by researchers in the learning disabilities field. These investigators have identified a number of approaches to the analysis of longitudinal data, and their work provides possible methodological models which could be applied by researchers interested in the developmental course of learning dis- abilities. Surely the complex interactions and transactions of learning disabled children and their environments over time should yield information and insights of real importance. Taken as a whole, then, it is clear that our understanding, even our de- scription of learning disabilities in different age groups and over time, is limited. Systematic study, particularly within a longi- tudinal design, is an important research need.

Sex Differences

Closely related to the question of age differences in the expression of learning disabilities is the question of possible sex related characteristics. Clinicians have for years recognized the discrepancies in referral rates of boys and girls for specialized services. Remedial classes and special education programs are usually filled with many more boys than girls, the ratios often varying from two to as high as ten to one. While characteristic of a number of clinical or special education conditions, the sex

discrepancy in incidence of identified learning disabled children is
striking. This is especially interesting given the recent finding
of equivalent incidence figures for underachievement of boys and
girls in an unreferred sample (Lambert & Sandoval, 1980).

Based on a comprehensive review of the literature on sex differences, Reid (1981) has identified a number of hypotheses to
explain the clinically documented ratios. These include differences
in the timing and expression of problem behaviors, differences in
the achievement areas affected, differences in salience of behavioral correlates, and differential patterns of teacher referrals.
It is a reasonable hypothesis, too, that sex linked differences in
psychological processing abilities may be related to learning disabilities. In this regard the work of Satz and of Knights (this
volume) and other neuropsychologists (Reitan, 1974; Rourke, 1975)
is especially pertinent. As noted by Satz and Morris (1980) in
their recent review of neuropsychological aspects of learning disabilities, there are a number of different subgroups contained
within the broad classification term, learning disabilities, these
subgroups differing on selected neuropsychological characteristics.
The interactions of possible sex-linked processing abilities and the
expression of learning disabilities is clearly an important area of
study.

From a somewhat different perspective it is important to emphasize that powerful socialization influences must be considered when
studying sex differences in incidence and kind of learning disabilities. This is particularly important given the transactional
approach proposed by Sameroff (1975) already referred to. The kinds
of expectancies adults have for acceptable and unacceptable behaviors and accomplishments may be potent influences on the nature of
children's behaviors. This perspective leads also to consideration
of a whole set of affective and motivational variables, including
attribution and expectation models. It is interesting to speculate
that as social roles change there may be differences in the sex-
linked nature of learning disabilities. The socialization changes
in the contemporary society, thus, provide an intriguing area of
study.

Whatever the basis for the disparate incidence of identification of boys and girls as learning disabled, the consequence has
been a confounded literature (Keogh et al., 1981). Said directly,
most generalizations about learning disabled children have been
drawn from samples of learning disabled boys. In the UCLA Marker
Project two of 408 studies included girls only; 50 of 408 were all
males; of the remaining studies which included both boys and girls,
the ratios of sexes varied from 4 to 1 to as high as 20 to 1. The
higher incidence of boys than girls in learning disabilities programs as well as in research samples continues to confuse findings

and generalizations. This is not just a problem for researchers and theorists, as the differential identification rate may carry signif- icant implications for practice (Johnson, 1975).

In summary, it seems safe to say that to date we know a good deal about learning disabled boys; we know very little about learn- ing disabled girls; and, our generalizations about learning disabled children are tenuous. The question of sex differences in learning disabilities is a topic which holds a high research priority.

Social and Cultural Influences

A third important area of needed research relates to culturally and economically different groups. Exclusionary criteria in many difinitions and in many research samples preclude identification and selection of possible learning disabled children in particular sub- groups. To illustrate, the definition of specific learning dis- abilities proposed in 1967 by the National Advisory Committee on Handicapped Chidren excludes children whose learning problems "... are due primarily to visual, hearing, or motor handicaps, to mental retardation, emotional disturbance, or to environmental disadvan- tage." The definition in PL 94-142, the Education for All Handi- capped Children Act of 1975, contains essentially the same language but adds an exclusionary statement relative to "... cultural, or economic disadvantage," as well as to environmental disadvantage.

It might be argued that if abilities are distributed normally within the population, we should expect to find learning disabil- ities represented in a number of subgroups now specifically ex- cluded by most definitions. It might be argued further that if learning disabilities are, in part at least, related to conditions of physical health, specifically to pre- and perinatal status, we might expect a higher incidence of learning disabled children from economically disadvantaged homes, where health services may be limited, than from groups where health care is more adequate. The transactions of biological and social conditions are well articu- lated by Sameroff and Chandler (1975) in their review of early risk conditions. Clearly there are powerful social, cultural, and eco- nomic factors which affect the development of learning disabilities. Given their importance it is rather ironic that social and cultural factors work to limit the selection and identification of children for intervention programs and for inclusion in research studies.

The learning disabilities diagnosis has somewhat facetiously been called a "middle class condition." Examination of the research literature does indeed suggest that the majority of samples have been limited to middle or upper socioeconomic status children; representatives from non-Anglo groups are conspicuous by their absence. The study of learning disabilities in a broader band of cultural, socioeconomic, or ethnic groups is obviously difficult and politically sensitive. Yet if we are to gain real understanding of the condition it is imperative to include representation from a broad social-cultural continuum. As already noted, we have drawn generalizations about learning disabled children on the basis of findings about learning disabled boys. By the same token we have talked of learning disabled children based on studies of middle socioeconomic children from the majority culture. A different perspective in our research efforts may very well lead to important insights which cannot be discerned from a restricted range of subjects.

A Broad Range of Research Variables

While discussion of research needs to this point has focused primarily on sample related issues, it is important to underscore the relatively limited number of topics which have received the systematic attention of researchers. The literature on documentation of child characteristics has proliferated, but the characteristics studied have for the most part been those contained within the traditional definitions, e.g., visual perception, auditory processing, fine and gross motor functions, attention, and language. As documented in the UCLA Marker Variable Project, cognitive and achievement tests make up the bulk of assessment data. These ability areas are important, to be sure. Yet, there is a striking lack of data on motivational or affective components or on correlates of learning disabilities.

The problem in part may be one of measurement, as we have few objective techniques to assess attitudes, attributions, and affective and motivational variables. By the same token we have few satisfactory ways to assess environments, be they home or schools. Some efforts have been made to capture the functional aspects of home environments which influence or which interact with children's abilities to produce growth. Wolf (1968), for example, described differences in homes in terms of achievement related variables such as emphasis on accomplishment, and parent involvement with their children's educational programs. This kind of analysis is particularly interesting given possible cultural differences in child rearing and family life. In this regard studies of high achieving Mexican-Americans are particularly instructive, as these may identify several family interactional variables which relate to academic and professional accomplishment. As noted by Price-Williams and Gallimore (1980), cultural processes affect the nature of children's modes of thinking, their aspirations, self-views, and problem-solving styles. In short, recognition of the complex multivariate nature of development leads to increased awareness of the importance of many extra-child factors as well as to consideration of a broad range of child characteristics.

CONCLUSIONS

From the perspective of a transactional model we would anticipate that growth is not a simple additive, linear accumulation of
skills, nor a strict biologically determined unfolding of matura-
tional potential. Discontinuity, not continuity, may be the rule
rather than the exception, especially for children with problems in
development. There are considerable data which demonstrate the weak
predictive power of single variables for subsequent development. On
a clinical level we all have had experiences with children who were
high risk for achievement problems because they had perceptual,
motoric, or attentional problems; yet, many of these children had
powerful compensating abilities, or were in home and school settings
which were facilitating and enhancing to growth. Explicit acknowl-
edgement of the broad range of within- and extra-child variables of
potential importance, and incorporation of multivariate strategies
may help identify the continuities and discontinuities which have
puzzled and frustrated us to date.

Overall our research efforts have been characterized by enthusiasm which is typical of early periods in any area of study.
Enthusiasm is necessary but not sufficient, however. Thanks to researchers in a variety of disciplines a number of theories and
models seem promising and testable when applied to learning disabilities. Data analytic strategies are increasingly sophisticated
and powerful. Years of clinical study have provided provocative and
rich hypotheses. The time seems right for systematic study of this
complex and intriguing topic.

References

Ackerman, P., Dykman, R. A., & Peters, J. E.   Learning disabled
     boys as adolescents. Journal of the American Academy of Child
     Psychiatry, 1977, 16, 293-313.
Adelman, H. S.   The not-so-specific learning disability population.
     Exceptional Children, 1971, 37, 528-533.
Balow, B., & Blumquist, M.   Young adults ten to fifteen years after
     severe reading disability. Elementary School Journal, 1965, 66,
     44-48.
Baltes, P. B., Reese, H. W., & Nesselroade, O. R.   Life span
     developmental psychology: Introduction to research methods.
     Monterey, California: Brook/Cole, 1977.
Education for all Handicapped Children Act. Public Law 94-142,
     94th Congress, S.6, November 29, 1975.
Gottesman, R. L.   Follow-up of learning disabled children.
     Learning Disability Quarterly, 1979, 2, 60-69.
Hechtman, L., Weiss, G., Finkelstein, W., & Benn, W.   Hyperactives
     as young adults: preliminary report. Canadian Medical Associa-
     tion Journal, 1976, 115, 625-630.

Huessy, H. R., & Gendron, R. A.   Prevalence of the so-called hyper-
    kinetic syndrome in public school children in Vermont. Acta
    Paedopsychiatry, 1970, 37, 243-248.
Huessy, H. R., Metoyer, M., & Townsend, M.   Eight to ten year
    followup of 84 children treated for behavioral disorder in rural
    Vermont. Acta Paedopsychiatry, 1974, 40, 230-235.
Johnson, V. M.   Salient features and sorting factors in diagnosis
    and classification of exceptional children. Peabody Journal of
    Education, 1975, 52, 142-144.
Kearsley, R. B.   Iatrogenic retardation: A syndrome of learned
    incompetence. In R. B. Kearsley and I. E. Sigel (Eds.), Infants
    at risk: Assessment of cognitive functioning. Hillsdale, New
    Jersey: Lawrence Erlbaum and Associates, 1979.
Keogh, B. K., Major, S. M., Omori, H., Gandara, P., & Reid, H. P.
    Proposed markers in learning disabilities research. Journal of
    Abnormal Child Psychology, 1980, 8, 21-31.
Keogh, B. K., Major, S. M., Reid, H. P., Gandara, P., & Omori, H.
    Marker variables: A search for comparability and generaliza-
    bility in the field of learning disabilities. Learning
    Disabilities Quarterly, 1978, 1, 5-11.
Keogh, B. K., Major-Kingsley, M., Omori-Gordon, H., & Reid, H. P.
    A system of marker variables for the field of learning
    disabilities. New York: Syracuse University Press, in press.
Lambert, N. M., & Sandoval, J.   The prevalence of learning
    disabilities in a sample of children considered hyperactive.
    Journal of Abnormal Child Psychology, 1980, 8, 33-50.
Major-Kingsley, S.   Outcome of learning disabilities: Inferences
    of the nature of the disability and intervening experiences upon
    adjustment in young adulthood. UCLA Graduate School of Educa-
    tion, in progress, 1981.
Minde, K., Weiss, G., & Mendelson, N.   A 5-year follow-up study
    of 91 hyperactive school children. American Academy of Child
    Psychiatry Journal, 1972, 11, 595-610.
National Advisory Committee on Handicapped Children. Conference
    sponsored by the Bureau of Education for the Handicapped, U.S.
    Office of Education, Washington, D.C., September, 1977.
Preston, R. C., & Yarington, D. J.   Status of fifty retarded
    readers eight years after reading clinic diagnosis. Journal of
    Reading, 1967, 11, 122-129.
Price-Williams, D., & Gallimore, R.   The cultural perspective.
    In B. Keogh (Ed.), Advance in special education: Volume 2,
    Perspectives on applications.
Rawson, M. B.   Developmental language disability: Adult accom-
    plishments of dyslexic boys. Baltimore: Johns Hopkins Press,
    1968.
Reid, H. P.   Sex differences in learning disabilities: A systems
    analysis. UCLA Graduate School of Education, in process, 1981.

Reitan, R. M.   Psychological effects of cerebral lesions in chil-
    dren of early school age. In R. M. Reitan and L. A. Davison
    (Eds.), Clinical neuropsychology: Current statistical appli-
    cations. Washington, D.C.: L. H. Winston and Sons, 1974.
Robinson, H. M., & Smith, H. K.   Reading clinic clients--Ten years
    after. The Elementary School Journal, 1962, 63, 22-27.
Rourke, B. P.   Brain-behavior relationship in children with
    learning disabilities: A research program. American Psycholo-
    gist, 1975, 30, 911-920.
Salvia, J., & Ysseldyke, J. E.   Assessment in special and remedial
    education. Boston: Houghton Mifflin, 1978.
Sameroff, A. J.   The etiology of cognitive competence: A systems
    perspective. In. R. B. Kearsley and I. E. Sigel (Eds.), Infants
    at risk: Assessment of cognitive functioning. Hillsdale, New
    Jersey: Lawrence Erlbaum & Associates, 1979.
Sameroff, A. J.   Early influences on development: Fact of fancy?
    Merrill-Palmer Quarterly, 1978, 21, 217-293.
Sameroff, A. J., & Chandler, M. J.   Reproductive risk and the
    continuum of caretaking casualty. In F. D. Horowitz (Ed.),
    Review of child development research, Vol. 4, Chicago:
    University of Chicago Press, 1975.
Satz, P., & Morris, R.   Learning disability subtypes: A review.
    In F. J. Pirozzolo and W. C. Wittrock (Eds.), Neuropsychology
    and cognitive processes in reading. New York: Academic Press,
    in press.
Werner, E. E.   Environmental interaction in minimal brain dys-
    functions. In H. Rie and E. Rie (Eds.), Handbook of minimal
    brain dysfunctions. New York: John Wiley, 1980.
Werner, E. E., Bierman, J. M., & French, F. E.   The children of
    Kauai: A longitudinal study from the prenatal period to age
    ten. Honolulu: University of Hawaii, 1971.
Wolf, R.   The measurement of environments. In N. H. MacGinitie
    and S. Ball (Eds.), Readings in psychological foundations of
    education. New York: McGraw-Hill, 1968.

DEFINITION OF LEARNING DISABILITIES: A CLINICAL APPROACH

Rosa A. Hagin, Ronnie Beecher, and Archie A. Silver

Fordham University School of Education and
New York University School of Medicine
Department of Psychiatry, New York University
School of Medicine
University of South Florida Medical School at Tampa

Defining learning disabilities is a thankless task.   Neverthe-less, it is necessary for the conduct of meaningful research, accurate diagnosis, and effective remediation.   Some of the current definitions have been criticized for their use of exclusionary cri-teria, questionable statistical procedures, and ambiguities in ter-minology.   It is helpful to examine the clinical implications of each of these criticisms.

Criticisms of Current Definitions

A common practice is to define learning disability as the fail-ure to learn well despite conventional educational opportunity, ade-quate intelligence, appropriate motivation, and normal sensory acuity (Eisenberg, 1967).   While it may be reasonable to attempt to differ-entiate learning disability from the results of social disadvantage, mental retardation, emotional disturbance, and sensory defects, the application of these exclusionary criteria in the cases of individ-ual children is not easy.   Because the problems of real children are often complex and multiply-determined, many questions can be raised. Does learning disability always occur independently of other handi-capping conditions?   Does poverty produce a different kind of learn-ing failure from affluence?   Can one separate emotional reactions to learning failure from the learning disabilities themselves?   Is it possible that the organizational and retrieval problems associated with learning disabilities can result in a lower score on intelli-gence tests?   The exclusionary criteria seem to be based on the doubtful assumption that having one kind of problem precludes any other.

Definitions of learning disability on the basis of a discrepancy between scores on intelligence and achievement tests (Bond & Tinker, 1967; Harris, 1970) have also drawn criticism. Regression effects and errors of measurement in the test scores themselves, as well as in the difference between any individual's scores on both tests, have been pointed out by several writers (McLeod, 1979; Rutter, 1978). McLeod has proposed the use of predicted values based on regression equations between aptitude and achievement tests as mathematically, psychometrically, and logically more justifiable. However, one might wonder about the clinical utility of such an approach, because it would require one to generate prediction tables for all tests, ages, and testing points in the population to be served.

The many variations in the description of learning disabilities in the professional literature do not produce conceptual clarity. Indeed, the ambiguities sometimes generated make the task of definition even more difficult. These variations result from a number of sources. The subjects of the studies may represent diverse age groups and, since it is likely that different factors contribute to learning failure at different ages, results can be contradictory. Sources of samples may also vary, with clinically-referred samples reflecting potential bias in terms of sex, economic status, and degree of pathology. Still another kind of variation may be introduced by the background of the investigator. Terminology, variables selected for study, experimental procedures, and emphases in interpretation of results may reflect the discipline in which the investigator was trained.

Guidelines for a Clinical Definition

While these criticisms do not exactly encourage one to attempt a definition of learning disabilities, they do suggest some guiding principles: (1) definitions should be based on multi-dimensional data, (2) definitions should provide for variations within individual subjects, (3) the age of subjects is significant in any formulation, (4) measurement and sampling considerations influence generalizations made about the nature of learning disabilities.

These guiding principles led to the selection of data from the school-based projects of the Learing Disorders Unit for a clinical approach to definition. These data were drawn from intact groups of children all born in 1971 who were scanned as part of the ongoing school-based preventive projects operated by the Learning Disorders Unit of the Department of Psychiatry of New York University Medical Center, in cooperation with Community School District II, Manhattan. These are clinical data collected in the natural setting of seven inner-city schools. Scanning procedures involve every child enrolled, in order to avoid the economic, sexual, and social biases that referral procedures may introduce. A sample of young children was chosen in order to minimize the effects of emotional reactions to learning failure, as well as the effects of teaching approaches upon the variables studied.

BACKGROUND ON THE STUDY

The Learning Disorders Unit is a multidisciplinary group with both service and research responsibilities. Since 1969 the Unit has operated school-based projects to locate children vulnerable to learning failure, to provide interdisciplinary clinical diagnosis, and to intervene educationally before failure has occurred. From 1976 to 1979 it was a model program funded by the Bureau of Education for the Handicapped. The preventive program was submitted to the Joint Dissemination Review Panel of the U.S. Office of Education and the National Institute of Education in 1979 where it was approved on the basis of data concerning educational impact, replicability, and cost effectiveness. It is now part of the National Diffusion Network of the U.S. Department of Education.

The core of this program is a scanning test SEARCH, which is administered to all children in participating schools before they enter first grade, and TEACH, a program of learning activities designed to mesh with the educational needs of vulnerable children as revealed by scanning and diagnosis (Hagen, Silver, & Kreeger, 1976; Silver & Hagen, 1976).

Our criteria for vulnerability were drawn from the features that our clinical studies have shown to distinguish children with learning disability from children with other clinical problems: basic immaturity with respect to the child's age and intelligence in aspects of the neurological, psychological, and perceptual examination dealing with spatial orientation and temporal organization. Clinical surveys have described these findings in detail (Silver & Hagin, 1960) and follow-up studies of children from this sample have traced the pervasive influence of temporal and spatial immaturities on aspects of perception and language and on personality development (Silver & Hagin, 1964).

When we moved from the clinic into our first school-based project in 1969, we found that the clinical criteria could be applied with the intact first grade sample from a regular public school (Hagin & Silver, 1977). These criteria correctly predicted for 98% of the children later found to have learning disabilities. In a sample of 168 children, there was one false positive (e.g., a child who, on examination was found to have the signs described above, but who was succeeding in learning on follow-up in third grade). There were nine false negatives (e.g., children who did not learn effectively although they had not demonstrated the clinical findings described above). Of this group of nine, two children were found to have emotional problems reactive to family disorganization. The remaining seven were bilingual children whose abstract language abilities were not equal to the conceptual demands of the reading comprehension test given in third grade.

Having surveyed the incidence and variety of problems in the natural setting of the school, the Learning Disorders Unit began what has become a long-term commitment to prevention through early identification, diagnosis, and intervention with learning disabilities. With wider dissemination of preventive programs, we found we could not continue to enjoy the luxury of intensive clinical study of all children. Our solution to meet the dilemma of cost, on one hand, and the need to preserve the cues for intervention available from intensive clinical study on the other, was to plan a two-step process: scanning and diagnosis.

The first step is to locate children who are vulnerable to learning failure. This is done by means of a twenty-minute individual test SEARCH (Silver, Hagin, & Beecher, 1978).

This test consists of ten components: three in visual perception (discrimination, recall, and visual-motor), two in auditory perception (discrimination and sequencing), two in intermodal skills (articulation and sound-symbol association), and three in body image (directionality, finger schema, and pencil grip). The ten components are utilized as a total SEARCH score to delineate vulnerability and as individual scores to yield a profile of assets and deficits to guide educational intervention.

While the first step, scanning, identifies the children who are vulnerable, the second step, diagnosis, attempts to find the reasons for their vulnerability. These multidisciplinary clinical studies consist of neuropsychiatric, psychological examinations, and developmental history from parents and school and hospital records. Diagnostic procedures, including ratings for the neurological and psychiatric examinations and diagnostic use of the Wechsler Preschool and Primary Scale of Intelligence have been described in previous publications (Hagin & Silver, 1977; Hagin, Silver, & Corwin, 1972; Silver & Hagin, 1972).

Results

Within the framework described above, this chapter will focus on one group of children, born in 1971. This group consists of children who had been enrolled in kindergarten (BD 1971) or entered early in first grade in seven elementary schools located in the lower east side or midtown Manhattan. A total of 650 children were

scanned with SEARCH by September, 1977. There were slightly more
boys (52%) than girls (48%) in the group. The ethnic census for
these schools showed that there were 11.5% Black, 36% Oriental, 34%
Hispanic, and 18.5% Other, including Caucasian children, enrolled.
In our BD 1971 sample, thirteen children were found to have insufficient English for valid testing and were dropped from this sample.
Otherwise, the group scanned represented the intact enrollment of
beginning first graders of the seven schools.

There were 192 or 29% of the group scanned who were found to be
vulnerable to learning disability. Of these, 103 (54%) were boys
and 89 (46%) were girls. These proportions closely approximate the
sex distribution in the total group.

By the end of the first grade the effects of mobility could be
seen in our sample. Of our original group, 494 or 76% remained enrolled in the seven project schools. As is often the case, the
children most in need of services are also the most mobile. By the
end of first grade, 124 (64%) of the children originally identified
as vulnerable remained enrolled in the project schools. The percent-
age of vulnerability (124 out of 494) becomes 25.4% with sex distribution relatively the same (60% boys 40% girls). The ethnic distribution in the vulnerable group closely paralleled the ethnic distribution of the seven schools: 9% Black, 30% Oriental, 41% Hispanic,
20% Other. On the WPPSI Full Scale IQ scores for the vulnerable
children ranged from 61 to 124, with a mean of 93.39 ± 12.43.

Diagnostic study enabled us to identify a number of subgroups
within the sample of vulnerable children:

Children with Specific Language Disabilities. These children
demonstrated the basic problems with spatial orientation and temper-
al organization, evidence that cerebral dominance for language had
not been firmly established, and immaturity in body image and right-
left discrimination. Psychological study showed that they possessed
adequate experiential background for school learning. No evidence
of structural defect of the central nervous system or of the peripheral sensory apparatus was seen on neuropsychiatric examination or
in the children's developmental histories. However, they had difficulty with associative learning tasks such as association of sounds
with their written symbols and with the Animal-House subtest of the
WPPSI. The origin of this kind of learning disability may be
familial, (e.g., the child's inherited pattern of development is one
in which language specialization develops more slowly than other
brain functions). In the BD 1971 group, 83 or 17% of the children
were diagnosed as specific language disabilities. Proportions of
boys and girls were not significantly different, with 58% and 42%
respectively diagnosed as specific language disabilities. Sam is an
example of the youngsters in this group:

The psychiatrist found Sam to be "a well developed boy
who was friendly yet well-controlled during the examination. He was self-confident to the point of presenting a macho image, which may mask subtle doubts and
concerns about his abilities. There were definite
immaturities in finger gnosis, left-right discrimination and spatial orientation. The remainder of the
neurological examination was within normal limits."

His functioning on the WPPSI placed within the average
range with a Full Scale score of 100, a Verbal IQ of
107, and a Performance IQ of 92. He was a slow delib-
erate worker. Verbal concepts were well developed and
ideas expressed easily, often with complex grammatical
patterns. High point in the record occurred with a
test of social judgment on which he earned a superior
score. The only significantly low point was with the
Animal-House subtest. Sam worked very slowly, naming
animals and colors as he worked. He did not learn the
code.

Children with Neurological Deviations. These children present
the common characteristics of problems with spatial orientation and
temporal organization, but in addition have deviations in one or
more areas of the neurological examination. This examination includes assessment of muscle tone, power, and synergy; gross and fine
motor coordination; cranial nerve functioning; maintenance of posture and equilibrium; deep, superficial, and pathological reflexes
as described in previous papers (Silver & Hagin, 1972). Rarely do
findings for children in this group point to focal brain damage and
rarely can specific causal factors be found in their histories.
Children with neurological deviations numbered 23 or 5% of the BD
1971 group. Of these 23 youngsters, 16 were boys and 7 were girls,
a proportion significantly different from the equal distribution of
boys and girls in the total sample.

Even within this small subgroup some variation is seen. On
the basis of neurological data this group can be subdivided according to three clusters of symptoms: children with attention
deficits, children with problems in motor coordination, and children
with deviations on the classical neurological examination.

Children with Attention Deficits. The six children (4 boys and
2 girls) in the attention-deficit group presented a restless, driven
quality in their behavior. Their quick, impulsive movements were
reflected in difficulties in many aspects of the examination. Neurological evaluation was difficult because of the uneven quality of
their responses, as illustrated in Yolanda's record:

In addition to extreme hyperactivity, Yolanda had
nystagmus on lateral gaze, difficulty with tandem
walking and fine motor control, extinction of tactile stimuli. She earned IQ scores in the low
average to borderline ranges on the WPPSI with subtest scores varying from a low of 4 to a high of 14.
The psychologist commented that she was an extremely
difficult child to test because of her poor attention
and impulse control. She had to be instructed repeatedly to sit down and to attend to the task at
hand. Enroute to the testing room she had to be
physically restrained from plunging down the stairs.
She said that she occasionally brought herself to
school, but her response to the Comprehension subtest did not reflect the type of judgment and impulse
control that are necessary for such independence.

Children with Motor Problems. Six children (4 girls and 2 boys)
demonstrated problems with motor control. In contrast to children
with attention deficits, this group responded easily in examination,
as illustrated in the summary on Jose:

Jose was described by the psychiatrist as sociable
and affectionate. He was verbal, despite some oral
inaccuracies in his speech persisting even at the
age of six years, five months. Few positive findings appeared on the neurological examination,
except for difficulties with fine motor coordination,
adventitious movements and convergence of the out-
stretched arms, errors in praxis and aspects of body
image such as finger gnosis and right-left discrimination. He was hypoactive. Jose's uncertain line
quality and body image distortions were seen in his
human figure drawing and his reproductions of some
simple geometric designs from SEARCH. On the WPPSI
his verbal score was average, while his performance
score fell somewhat below average. The psychologist
commented that, while he was initially active, this
restlessness quickly subsided as he began to enjoy
the test materials. Good verbal abilities were
demonstrated in the superior scores on the Similarities and Comprehension subtests of the WPPSI, although there were some gaps in English vocabulary
and information. Jose worked very slowly on the
Animal-House subtest. He also had difficulty keeping
his place in the rows on the board and remembering to
refer to the code sample rather than his own incorrect responses.

Children with Neurological Findings. Eleven children (only 2 of whom were girls) demonstrated positive findings on the classical neurological examination. Children in this subgroup generally require provision for overlearning in any teaching plans, not only because of the wide range of symptoms, but also because of their predisposition to anxiety. The common characteristic among these children is their variability, both in terms of the individuals comprising the subgroups and in the day-to-day functioning of each child. Al is an example of this subgroup:

Al showed positive findings in the neurologist's examination of cranial nerves, reflexes, and cerebellar signs. Muscle tone was decreased, and myclonic activity was present. He had difficulty with patterned motor behavior and fine motor coordination. Vasomotor and pupilary abnormalities suggested dysfunction of the autonomic nervous system. Extinction of tactile stimuli and errors in finger gnosis were noted.

Al's relationship to the examiner was shy and inhibited. He seemed aware of his problems, fussy, insecure, and easily overwhelmed. As his anxiety increased, he attempted to avoid difficulties by responding in a humorous manner.

Scores on the WPPSI placed in the average range for the verbal scale and in the dull normal range for the performance scale. He earned superior scores with items tapping social awareness and vocabulary. He had difficulty with the sequencing of meaningful material and also with the associative learning task. The psychologist commented on his sluggish articulation, clumsiness with fine motor control, and awkward gait. Al was strikingly aware of his problems and tended to respond defensively when he felt overwhelmed.

Children with General Immaturity. Within the sample there was a group of children with nonspecific learning problems. This group was characterized by generalized immaturity. With them there was no clinical or historical evidence of structural defect of the central nervous system, but there was slowness reported in reaching developmental landmarks. The uniformly low curve of maturation was apparent in their physical appearance.

They were found to be small in size. In gross and fine motor development, in language, and in social awareness, they appeared to be younger than their chronological ages. Although these children function below average on psychological study, we did not find indications of any currently described syndrome of mental retardation.

Three children (0.6% of the total sample) comprise this group. An example is Nora:

The psychiatrist described her as "a small, delicately built, beautiful six-year-old who seemed younger than the other first graders. She was shy and somewhat inhibited at first, but this faded rapidly into a candid expression of her feelings. She discussed concerns about her difficulty in reading; this seemed to be a sensitive area for her.

The only findings on the neurological examination were immaturity in left-right orientation, finger gnosis, and temporal sequencing. The remainder of the neurological examination was within normal limits."

Nora earned a Full Scale of 81, a Verbal Scale of 84, and a Performance Scale of 82. Subtest scores ranged from 5 to 9 with no significant deviations from the mean scaled score of 7. In the testing room rapport was easily established with Nora initiating conversation. Her response to difficult items was to withdraw, although she would try items with some encouragement. She tended to respond in a global manner, like a younger child. Some verbal concepts were well-developed, yet there were gaps in her expressive vocabulary.

Other Children Identified as Vulnerable. Within the group of children also identified as vulnerable were 15 youngsters whose neurological examinations were found to be within normal limits. Of these six children (5 boys and 1 girl) demonstrated on psychiatric examination emotional problems serious enough to impair learning. Some of these children were reacting to pathological conditions within the family and the resulting disruption in the home. Others presented severe neurotic and behavior problems reflected in every aspect of their lives, including school learning.

There were four children for whom diagnosis was deferred. Of these, two children were found to have mild hearing losses which needed to be evaluated and, if necessary, treated. In addition, there were two children, newly arrived in the United States who were culturally different youngsters, for whom scanning and diagnostic measures were regarded as inappropriate.

Finally, there were five children, identified as vulnerable on scanning, who proved to be normal learners on diagnosis. This represents a 1% rate of false positives on our prediction measure; the rate of false negatives was found at the end-of-the-year achievement testing to be 8.5%.

Discussion

This preventive model of the Learning Disorders Unit has provided the opportunity to study in their natural setting intact
groups of young children as they began their academic work at
school. Methods have been developed to identify children who were
vulnerable to learning failure before they failed. The distress
caused by the high mobility rate in some of our samples was more
than balanced by the data obtained on sex ratios.

In 1969 it was noticed that there were equal numbers of boys
and girls in the vulnerable groups. This was thought to be an arti-
fact because the literature reported consistently that learning disability was six-, seven-, or ten-times more frequent in boys than in
girls. Yet relatively equal numbers of boys and girls continued to
appear in our samples of vulnerable five- and six-year-olds. BD
1971 was not unlike BD 1963 or BD 1973 in this respect. There was,
however, one diagnostic subgroup in which the boy-girl ratio
appeared to be significantly different: the group with deviations
on the neurological examination. Pediatric studies of newborns may
shed some light on this finding. In general, such studies show that
male babies tend to be larger and more vulnerable physically. It
may be, therefore, that they are also more at risk for neurological
insult at birth.

On the other hand, the reasons for the discordant referral
rates of boys and girls are not clear from our data. It is, however,
tempting to speculate on the impact of parental expectations for
boys, or male abrasiveness versus teacher irritability, or differential coping styles among boys and girls.

As for the original problem of the definition of learning disability, the concept of immaturity in spatial orientation and temporal organization is offered as an attempt to unify the diverse
symptoms that have been described. With intact groups of five- and
six-years, a scanning measure based on this concept correctly predicted learning disability with 1% false positives and 8.5% false
negatives in our sample.

Diagnostic studies have shown that, even at this early age, the
children thus identified are not a homogeneous group. It is this
multidimensionality that invites confusion, frustration and failure
of replication in research on learning disabilities. One cannot
assume that the defects associated with learning failure in one
child are the same as those in any other child who fails to learn.
In reality, within the group of children designated as learning disabilities, there are marked individual differences in functioning
and in assets and deficits. The importance of these variations is
that, in any study of large numbers of children, these individual

differences may cancel each other out. This balancing effect may
cause an investigator to dismiss as unimportant some variables that
may be crucial to some of the children in the sample.

While the complexity of learning disabilities makes multidi-
mensional study advisable, the developmental nature of the condition
gives strong support for a longitudinal approach as well. The ability of the human organism to mature and to compensate for functions
that have lagged behind the child's general development may produce
misleading conclusions when one depends upon cross-sectional data
alone.

For example, Satz, Taylor, Friel, and Fletcher (1978) found
that neuro-perceptual deficits identified in kindergarten predicted
reading achievement at the end of second grade, despite the fact
that in their tests the deficits had largely been overcome by that
age. Satz et al. (1978) found that, while perceptual deficits were
more characteristic of learning disabled children when they were
young, difficulty with language skills was more characteristic of
the same children when they reached the upper elementary grades.

Other studies using multivariate approaches have shown the complexities in predicting and diagnosing learning disabilities (Bell &
Aftanas, 1972; Jansky & de Hirsch, 1972). The investigator is frequently left with a list of characteristics which may be associated
with learning disabilities in some children, but may not be present
in other children. Which variables are significant may depend upon
the age of the children, the compensatory resources (cognitive,
emotional, motivational) and the environmental supports available to
them. It is the responsibility of the investigator to utilize these
often diffuse data to build coherent theories of learning disabilities to account for the phenomena their research may have highlighted.

It may be that no one definition can encompass the heterogeneous group of children who experience learning problems. As continuing investigations refine models within the broad category of
learning disabilities, it may be possible to delineate special characteristics that set these children apart from their achieving
classmates. It is proposed that, for the five- and six-year-old,
problems in spatial orientation and temporal organization are a useful identifying variable to signal the need for more specific diagnosis and intervention. As the natural history of learning disorders is better understood, other characteristics may be found to
serve this identifying role with other age groups.

It may also be that no one definition can serve all purposes
equally well and that different definitions for different purposes
will emerge. For example, a general, more inclusive operational

definition may be more serviceable for the clinician and the program
administrator to use in planning preventive and remedial programs.
In contrast, increasingly precise models of learning disability
would be required for investigators in order to facilitate replication studies and to encourage synthesis of research in our often
diffuse field. The research models should provide for the age-
specific, developmental, and ecological variations in the samples
under study. It is suggested that investigators abandon the search
for a single, all-purpose definition. Children and youth of this
heterogeneous field will be best served by definitions which are
specific to the purposes for which they will be used.

References

Bell, A. E., & Aftanas, M. S.   Some correlates of reading retarda-
    tion. Perceptual and Motor Skills, 1972, 35, 659-667.
Bond, G. L., & Tinker, M. A.   Reading difficulties: Their diagnosis
    and correation. New York: Appleton-Century-Crofts, 1967.
Eisenberg, L.   Clinical considerations in the psychiatric evalua-
    tion of intelligence. In J. Zubin and G. A. Jervis (Eds.),
    Psychopathology of mental development. New York: Grune and
    Stratton, 1967.
Hagin, R. A., & Silver, A. A.   Learning disability: Definition,
    diagnosis, and prevention. New York University Education
    Quarterly, 1977, 8, 9-16.
Hagin, R. A., Silver, A. A., & Corwin, C. G.   Clinical diagnostic
    use of the WPPSI in predicting learning disabilities in grade
    one. Journal of Special Education, 1972, 5, 221-232.
Hagin, R. A., Silver, A. A., & Kreeger, H. TEACH: Learning tasks
    for the prevention of learning disability. New York: Walker
    Educational Book Corporation, 1976.
Harris, A. J.   How to increase reading ability (5th ed.).   New
    York: David McKay, 1970.
Jansky, J., & de Hirsch, K.   Preventing reading failure. New York:
    Harper and Row, 1972.
McLeod, J.   Educational underachievement: Toward a defensible
    psychometric definition. Journal of Learning Disabilities,
    1979, 12, 322-330.
Rutter, M.   Prevalence and types of dyslexia. In A. L. Benton and
    D. Pearl (Eds.), Dyslexia: An appraisal of current knowledge.
    New York: Oxford University Press, 1978.
Satz, P., Taylor, H. G., Friel, J., & Fletcher, J.   Some develop-
    mental and predictive precursors of reading disabilities: A six
    year follow-up. In A. L. Benton and D. Pearl (Eds.), Dyslexia:
    An appraisal of current knowledge. New York: Oxford University
    Press, 1978.

Silver, A. A., & Hagin, R. A.   Search: A scanning instrument for the identification of potential learning disability. New York: Walker Educational Book Corporation, 1976.

Silver, A. A., & Hagin, R. A.   Profile of a first grade: A basis for preventive psychiatry. Journal of the Academy of Child Psychiatry, 1972, 11, 645-674.

Silver, A. A., & Hagin, R. A.   Specific reading disability: Follow-up studies. American Journal of Orthopsychiatry, 1964, 34, 95-102.

Silver, A. A., & Hagin, R. A.   Specific reading disability: Delineation of the syndrome and relationship to cerebral dominance. Comprehensive Psychiatry, 1960, 1, 126-134.

Silver, A. A., Hagin, R. A., & Beecher, R.   Scanning, diagnosis and intervention in the prevention of reading disability. Journal of Learning Disabilities, 1978, 11, 437-449.

PART II

Cognitive Processes

The following three chapters each propose somewhat different
approaches to the investigation of learning disability. The
chapters by Barclay and Hagen and by Das, Snart, and Mulcahy suggest
that the problems encountered by many learning disabled children
pertain to the use of appropriate strategies and their existing
knowledge. These two chapters emphasize the ability to plan as a
major aspect of the difficulties of many children with learning
problems. Whereas the first two chapters discuss the broad aspects
of strategies, planning and the processing of information, Torgesen
takes a much more specific approach in an examination of performance
on a single task. He suggests that the processing deficit of a
specific subgroup of learning disabled children (those with short
term memory problems) is a "structurally based" one. On the surface, this view appears to be contradictory to the planning-strategy
emphasis taken by Barclay and Hagen as well as by Das, Snart, and
Mulcahy. Torgesen does present some convincing data to support his
suggestion. In doing so, he compels us to consider carefully the
procedures by which a structurally based deficit can be clearly differentiated from one which is strategy based.

The authors of these three chapters emphasize very strongly
the need to examine subgroups of learning disabled children; to
determine task requirements by task analysis, to ascertain the
strategies and the knowledge used by the children, and then to
ultimately relate this information to instruction.

THE DEVELOPMENT OF MEDIATED BEHAVIOR IN CHILDREN: AN ALTERNATIVE
VIEW OF LEARNING DISABILITIES[1]

Craig R. Barclay and John W. Hagen

Department of Psychology
University of Michigan, Ann Arbor, Michigan

The problems of identifying and educating learning disabled
children are attracting increasing amounts of attention from developmental psychologists interested in cognitive processes. Queries
as to why disabled children fail on certain academic tasks provide
a common ground of concern for researchers and educators alike.
Perhaps the single most important issue is whether the poor school
performance of the disabled child reflects an inability to achieve
or a lack of task appropriate skill which could be acquired through
instruction. If taken from a cognitive-developmental perspective,
this issue is reflected in the extensive work which describes the
child's growing independence from immediate environmental controls
to the thoughtful regulation of behavior.

Much of the research on cognitive development has focused on
the apparent shift from the non-mediated, concrete thinking of the
preschooler to the mediated, abstract thinking of the older child.
Luria (1976), citing Vygotsky's (1934/1962) earlier work, describes
this shift as "... the transition from unmediated sensory reflection
to mediated, rational thinking" (p. 50). Mental development is
portrayed as a change in the content and form of cognitive activity;
the child's thoughts are no longer wholly situation specific but
represented more in the form of taxonomic categories. It is argued
that this transition is the product of social and cultural events in
the child's life, such as entry into school. Further, the higher-
order representations associated with mediated behavior are based on

----
[1]The authors were supported in part by Grant T32 HD07109-04 of the
United States Public Health Service (NICHD).

the development of a symbolic, language system (Vygotsky, 1934/1962),
or as Piaget (1971, 1976) argues, the child develops a semiotic
function. Language accounts, in part, for the generativity and generalizability of the child's behavior.

This rendering of cognitive development parallels that emerging
from more recent work on verbal mediation (e.g., Flavell, 1970; Hagen
& Stanovich, 1977) and organization (cf. Smiley & Brown, 1979, for a
discussion of the thematic-taxonomic shift). In brief, most children
develop the capability to govern behavior through language and represent environmental features in superordinate categories, e.g., living
versus non-living things, animals versus furniture. It appears that
many learning disabled youngsters experience great difficulty in the
implementation and regulation of behavior through symbolic systems
and related language skills.

The purpose of this paper is to review selectively the research
supporting the notion that the child's memory development results
from a growing ability to mediate behavior verbally. The work on
memory development is germaine to the arguments presented here since
an understanding of adaptation is inseparable from the issues of the
storage and use of past as well as present experience. More importantly, the mediation literature provides a frame of reference for
understanding the nature of many learning disabilities; in particular, those disabilities correlated with inactive verbal processing
during the acquisition of new information (e.g., Torgesen, 1977a).
It appears that many disabled children fail to process, spontaneously
and intentionally, to-be-learned material.

This paper is organized in four sections. In section one,
learning disabilities are described in terms of deficiencies in the
use of certain problem solving behaviors instead of central or peripheral structural deficits. A selective review of the literature
dealing with the development of strategies and knowledge is presented
in the second section. This is followed in section three by a discussion of our own work which focuses on the issues of strategy use
and metamemory in mentally retarded and learning disabled children.
In the final section, questions and issues for future research in
learning disabilities are addressed.

I. LEARNING DISABILITIES -- A DEFICIENCY VIEW

The basic problem faced by the learning disabled child is that
of acquiring knowledge through different academic tasks like reading.
The observation that these children amass large amounts of information from sources other than school-type exercises argues against
the view that they are incapable of efficient learning. In fact,
many current definitions of learning disabilities include the notions

that these children are of normal intelligence and not mentally retarded in the sense of an identifiable brain or neurological deficit (see also, Hagen, Barclay, & Schwethelm, 1981; Hallahan & Cruickshank, 1973). A learning disability is not characterized by a general intellectual deficit; instead, the disabled child tends to have a specific problem, e.g., an unorganized study routine, which may affect performance on a wide range of school tasks.

If it is the case that many disabled children are not necessarily neurologically impaired, then their subaverage academic achievement might be due to deficiencies in the use of relevant problem solving skills or strategies. The argument proposed here is that these children are strategy deficient; therefore, appropriate intervention procedures should enhance performance. Evidence from a number of literatures supports this argument, i.e., in learning disabilities -- Bauer (1977, 1979) and Torgesen (1977a, 1977b); in mental retardation -- Brown (1974), Brown and Barclay (1976) and, Campione and Brown (1977); in the development of "normal" children -- Flavell (1970) and Hagen (1971).

The perspective of the learning disabled presented here is similar to that taken of the young, non-disabled child who fails initially at some task, but, following a brief training period shows significant improvement. The point is that if the child were mediationally deficient (Reese, 1962), no amount of training would be helpful.

This "production deficiency" hypothesis is a familiar one. It assumes that children who do not spontaneously produce an effective mediational routine can be instructed to do so with a minimum amount of effort (Flavell, Beach, & Chinsky, 1966; see also, Paris, 1978, for a detailed discussion). Children who are production deficient may not have the needed skill available to them to solve a given problem, or simply do not think to use what they already know since test conditions do not facilitate the access to existing knowledge, or lack the intention to be strategic. Instructing children in strategy use addresses these important issues in the following ways. First, instruction guarantees that the child has an effective problem solving skill available which allows for subsequent investigations of the conditions which lead to the maintenance and generalization of what was taught. Second, through instruction, the child may be informed about the causative relationship between strategy use and enhanced memory performance or that a specific mnemonic is relevant to solving the problem at hand. Under such conditions it may be possible to study the roles played by knowledge and intention in children's mnemonic efforts. Also, from a more general perspective, if the child's performance improves following instruction, then structural dysfunctions cannot wholly account for memory performance.

The research cited in the next section supports the production deficiency explanation of poor memory performance. An attempt is made to show that young children, including many learning disabled and mentally retarded youngsters, are deficient in the use of strategies and knowledge which can be acquired through training.

II. THE DEVELOPMENT OF STRATEGIES AND KNOWLEDGE

Any description of development is concerned first with relating behavior change to age; thereby discovering when some behavior occurs in the child's life. Such a description provides the basis for theories about the course and sequencing of developmental events (e.g., Piaget, 1952). Once accurate description is achieved, a second concern is to determine the causes of behavior change over time. Each of these concerns reflects instances of the scientific enterprise -- one being correlational and the other manipulative (Cronbach, 1957). To date, most developmental research is descriptive, except for those cases in which some direct intervention occurs.

In memory development, questions about why the child's memory improves have focused mainly on possible capacity, strategy, and knowledge differences which covary with age (Barclay, 1981). Capacity explanations of memory development have taken two forms. One form posits that capacity refers to the architectual features of the information processing system; for instance, short- (STM) and long-term (LTM) memory (cf. Campione & Brown, 1977, for a review of this terminology). Changes in memory are assumed to reflect growth in such features. This position, however, is inconsistent with both theoretical and empirical arguments. It is assumed here that the architecture of the memory system is invariant, probably after about age 5 (Dempster, 1977, cited by Chi, 1978); that is, it does not change significantly over the life span. What does appear to develop are control processes which regulate the performance capacity of STM (Chi, 1978; Miller, 1956; Simon, 1974).

The other form of the capacity view of memory development hypothesizes that the "hardware" is fixed and development results from the child's growing ability to allocate attentional and strategic resources. An example comes from Gibson's (1969) work where it is hypothesized that the development of attention, as a cognitive process, results from the person's increased sensitivity to the affordance inherent in the environment. Changes in attention are tied closely to the recognition of essential environmental features. Thus, attention to relevant environmental features accounts, in part, for the increasingly efficient processing of information.

The study of strategic factors in the ontogeny of memory is associated with the development of mnemonic usage, typically in reference to language development. A mnemonic is any behavioral activity which aids memory. If a cognitive perspective is taken, mnemonics are strategies; that is, generalizable rules applied in a purposeful way to store and retrieve information. Mnemonic development is seen as the acquisition of strategies, e.g., rehearsal, which meet the demands of a myriad of tasks encountered in everyday life -- from remembering rote material to the semantic integration of large chunks of information. The extensive literature on mnemonic development is relevant here since it tends to show that young children, the mentally retarded, and learning disabled are comparable on the level of being production deficient with respect to strategy use (e.g., Brown, 1974; Flavell, 1970; Torgesen, 1977a).

An early view of the young child held that behavior was under the direct control of environmental stimuli which elicited and reinforced responding. In fact, it was thought that young children were incapable of controlling their own behavior since they were mediationally deficient (Reese, 1962). The mediation deficiency hypothesis stated that even if the child produced an appropriate strategy, performance would remain unchanged and relatively low. However, beginning with the seminal work of Kendler and Kendler (1962), evidence was gathered that described an apparent shift in the way the child solved problems. While the young child (three to six years old) learned to respond on the basis of a reinforcement history associated with the perceptual features of the task, the older child's (seven to ten years old) performance was explained as mediational in character (cf. Smiley & Weir, 1966, for clarification). Even though the Kendler's work supported a mediational view, two unanswered questions remained. First, what was the nature of the mediator used by the older child? Second, could the younger child use a mediator if prompted to do so?

The research reported by Flavell and his colleagues (Flavell et al., 1966; Keeney, Cannizzo, & Flavell, 1967) dealt directly with these questions. In the study by Flavell et al. (1966) it was shown that the likely mediator was language. The findings which supported this conclusion were that, given an ordered recall task, older children (Grades 2 and 5) used a verbal rehearsal strategy whereas younger children (kindergarteners) did not and the use of language (i.e., verbal rehearsal) was correlated positively with amount recalled. In the followup study (Keeney et al., 1967), first graders, who did not spontaneously produce a verbal strategy given a recall task, were taught to rehearse. In comparison to the appropriate controls, it was demonstrated that verbal rehearsal was related functionally to memory.

In the Keeney et al. (1967) study it was shown that children who did not spontaneously produce an effective mnemonic were trained easily to rehearse. The use of rehearsal resulted in improved recall -- approximately equal to that of spontaneous producers -- relative to a no training control (however, see Bebko, 1979). One interesting anecdotal finding was that of the 17 trained children, ten abandoned the rehearsal strategy when no longer prompted to use it. The authors suggested that those children who stopped rehearsing may not have been aware of the cause-effect relationship between strategy use and memory. That is, while training provided the necessary strategy it did not instantiate sufficient knowledge to lead to the unprompted maintenance of what was learned -- the trained children were still knowledge deficient.

Work reported by Kennedy and Miller (1976) tested the notion that production deficient children may be knowledge deficient. They replicated the Keeney et al. (1967) findings by training nonproducers to rehearse with the addition that half of the instructed children were given feedback regarding the beneficial effects of rehearsal. Not surprisingly, children in the feedback group continued to re-hearse during an unprompted maintenance test; whereas, the children in a no-feedback control group abandoned the instructed routine (see also, Asarnow & Meichenbaum, 1979; Gelabert, Torgesen, Dice, & Murphy, 1980; Paris, 1978). It seems then, that strategy training plus knowledge of the functional relationship between mnemonic use and memory are required before children continue to employ an effec-tive study routine without prompting.

Perhaps the major limitation to the findings reported thus far is that while several investigations have demonstrated the impor-tance of training specific mnemonics to "normal" children (e.g., Brown, 1975; Hagen & Stanovich, 1977; Flavell, 1970), the mentally retarded (e.g., Brown, 1974) and learning disabled (e.g., Bauer, 1977; Newman & Hagen, 1981; Torgesen & Goldman, 1977), few studies have reported on the generalization of a learned strategy to a new situation. Notable exceptions to this limitation, however, are found in the work of A. L. Brown and her associates (e.g., Brown, 1978; Brown & Barclay, 1976; Brown, Campione, & Barclay, 1979; Brown, Campione, & Murphy, 1976). In much of this research the child is informed that generalization is one important purpose of instruction. Also, the training procedures are designed to maximize their usefulness over problems of similar types (Campione & Brown, 1977). That is, transfer probes represent instances of "far gener-alization" since the process demands over tasks are held constant while the problem formats are changed, e.g., training a stop-and-check self-testing strategy is effective for acquiring rote as well as meaningful material. Stated differently, the important similar-ity between the instruction and generalization tasks is that the strategy trained is applicable for solving both types of problems.

The research cited here points to a multifaceted explanation of poor performance and the transition from production deficiency to the generalization of acquired skills. It seems that the conditions which maximize the acquisition, selection, and use of appropriate mnemonic skills are associated, in part, with how much children know about what they are taught. This knowledge not only includes the static, declarative features of some well learned routine but also a more general, dynamic feature. That is, a procedural knowledge typified by a sensitivity to when and why some strategic behavior improves performance (cf. Greeno, 1981; Resnick, 1981). This procedural knowledge may also lead to the intentional use of strategies. Unfortunately, to our knowledge, no evidence is available on the role intention may play in mnemonic usage, other than descriptions of certain children being "inactive" when asked to remember (e.g., Torgesen, 1981).

In the memory literature, knowledge is referred to as "metamemory" (cf. Flavell, 1970; Flavell & Wellman, 1977) as one form of "metacognition." The term metacognition is used in different ways depending on the domain to which it applied, e.g., language, social situations, or memory. Oftentimes this multiple usage leads to ambiguity and confusion. In general, metacognition refers to the knowledge a person has about cognition including an awareness of architectural features and processing capabilities. The relevance of metamemory for understanding learning disabilities is that the disabled are not only deficient in the use of certain mnemonics but they also may be unaware of the factors which affect their memory performance (cf. Torgesen, 1977b).

Flavell and Wellman (1977) propose a taxonomy which includes two categories of metamemory. One category is the child's increasing sensitivity to situations which demand effort or mnemonic activity to store and retrieve information (see also Kahneman, 1973). That is, the child becomes aware that certain types of problems require purposeful behavior to perform well. The other metamemory category is variables, or those factors which interact to affect the outcome of one's memory efforts. These variables are the person's knowledge of the (a) traits and states of memory, (b) task characteristics and demands, and (c) storage and retrieval strategies.

A knowledge of memory traits refers to the child's awareness of static memory features. For example, knowledge of the structural features of the memory system (Atkinson & Shiffrin, 1968; Campione & Brown, 1977). The complementary aspect of knowledge about traits is the child's knowledge of memory states; that is, the monitoring of the current contents of memory. Theoretically, memory monitoring is the process through which children recognize how much they already know and how much they need to know to meet the demands of a given task. Presumably, accurate monitoring allows the child to know when

the information is either learned or additional study is required
before indicating recall readiness (cf. Flavell, Friedrichs, & Hoyt,
1970).

Variables knowledge also includes the child's growing awareness
of different task characteristics and demands. Task characteristics
are the observable elements of the problem; that is, the amount and
kind of information to be learned. Instances of task characteristics
are the number of "chunks" to be remembered, mode of presentation,
e.g., written versus oral presentation, and nonsense or meaningful
material.

The other feature of the task variable is the child's knowledge
of demands. Task demands, unlike characteristics, are determined in
large part by the test requirements; typically the material must be
recalled, recognized, or reconstituted. Developmentally, these
three types of task demands can be ranked in terms of their relative
sensitivity to age related differences. Minimum age effects usually
are found given a recognition task (Brown, 1973; Brown & Scott,
1971) unless the information taps the child's knowledge base (e.g.,
Meyers & Perlmutter, 1978; also see Mandler & Robinson, 1978).
Unlike recognition, recall tasks reveal large age differences in
that older youngsters who use appropriate mnemonics outperform
younger nonstrategic children. Developmental trends are also found
on reconstruction tasks; however, the age differences are less pronounced relative to recall since the relevant items are available
during testing and associations among items can provide a structure
which enhances the children's performance.

The third type of "variables knowledge" proposed by Flavell and
Wellman (1977) is strategy, i.e., the child's knowledge of those
storage and retrieval mechanisms which can be used to meet the
demands of different kinds of problems. Interestingly, the hierar-
chy of tasks mentioned above results from the factor that recognition problems typically demand the least amount of strategic intervention, whereas, recall demands the most (Brown, 1975). The literature tends to support the view that with age, children not only
use effective encoding strategies, e.g., rehearsal, and decoding
strategies, e.g., retrieval cues (Kobasagawa, 1977), but they also
acquire knowledge of when and where to apply mnemonics together
with an awareness of why active information processing is needed
(e.g., Brown, 1975; Kreutzer, Leonard, & Flavell, 1975). Also, some
evidence suggests that through experience children become sensitive
to the ways in which the person, task, and strategy variables interact (e.g., Wellman, 1979).

In the next section a selected review of research is presented.
The focus of this review is on the evidence that strategies and
knowledge are two important components of effective problem solving.

III.   SELECTED STUDIES OF STRATEGY USE AND KNOWLEDGE

The purpose here is to review our work on strategy use and knowledge in normal (e.g., Barclay, 1979, 1981; Hagen, 1972; Hagen & Hale, 1973; Hagen, Jongeward, & Kail, 1975; Hagen & Stanovich, 1977) and mentally retarded children (e.g., Brown & Barclay, 1976; Brown, Campione, & Barclay, 1979; Buyer & Hagen, 1981; Hagen et al., 1981; Hagen & Huntsman, 1971; Hagen, Streeter, & Raker, 1974; Landau & Hagen, 1974; Newman & Hagen, 1981). The research reported is selective and the reader is referred to the papers cited above for a more complete overview.

Two types of related tasks were used in most of this research: serial recall and Hagen's selective attention task (cf. Hagen, 1967; Maccoby & Hagen, 1965). In serial recall, children are presented a series of items, usually pictures of common objects like dog, flag, hat, to be remembered in a predetermined order. One variation of this problem type was probed recall. In this task, a number of picture cards were first presented one at a time and arranged in a horizontal display. Each card was placed face down once the child looked at it. After the cards in a series were presented, a cue card was shown and the child was asked to point to the card in the display that matched the cue. The important feature of this task was that performance over trials was measured by the total number of items recalled as well as the amount of information remembered at each serial position. The initial positions of the resulting serial position curves are referred to as the primary portion, and the end positions, the recency portion. Theoretically, memory for items at the recency and primary portions reflect recall from short- and long-term memory, respectively (cf. Atkinson & Shiffrin, 1968; Ellis, 1970).

The selective attention task was a modified version of the serial recall problem and included an estimate of both central and incidental performance. Here, children were presented cards with two pictures per card, e.g., animal and household object, and instructed to remember one (central) of the items, e.g., animal. The children's memory for these pictures yielded a total central recall score. After a sequence of recall trials, two sets of cards were presented; on one were the central items, and on the other, the incidental ones. The children's task was to pair the central and incidental pictures that had appeared together on the same card during the initial serial recall trials.

In the early studies using serial recall, two questions were asked: (1) Does merely labeling the items to be remembered enhance recall? (2) If so, do labeling effects covary with age? These questions were tied theoretically to the issue of the verbal mediation of memory (cf. Flavell et al., 1966; Reese, 1962).

An initial study was reported by Hagen and Kingsley (1968). The children tested were four-, six-, seven-, eight-, and ten-years-old. In one condition, half of the children at each age were asked to name overtly the pictures while for the other half of the children in a control group, labeling was not required. The results indicated, not surprisingly, that recall improved with age. However, the effects of verbalizing interacted with age in a complex manner. Labeling facilitated recall at the intermediate ages but not for the youngest and oldest children. Analyses of recall by serial position revealed that for the oldest youngsters, verbalizing caused a decrement in primacy performance. Further, labeling had a beneficial effect on recall for all ages at the recency portions of the serial position curves. Apparently, the effects of overt verbalization are dependent upon the child's development level, i.e., the amount and quality of their language use.

Followup studies focused on the role verbal rehearsal might play in memory. Rehearsal can be viewed as any activity which keeps information alive in short-term memory and mediates the transfer of that material to a long-term store. Also, rehearsal differs from simple labeling in that multiple pieces of information are verbalized instead of single picture names.

In an experiment reported by Kingsley and Hagen (1969), five-year-olds were instructed to rehearse cumulatively the names of different animals. These children were selected since they failed to produce a verbal rehearsal strategy on their own (cf. Flavell et al., 1966; Keeney et al., 1967). Those children taught to rehearse cumulatively remembered more items than children instructed to label the pictures. Further, recall of primacy information was significantly elevated for the children who rehearsed, and no differences were found between the rehearsal and label groups on recency items. Therefore, instructing children who do not spontaneously produce an appropriate verbal strategy improved their recall.

These results were replicated and extended by Hagen, Hargrave, and Ross (1973). The important feature of this study was that a condition was added in which rehearsal was taught but required only during the presentation of the pictures. No prompting to rehearse was given when recall errors were made. The results indicated that recall declined for five- and seven-year-olds. Also, the performance of children who initially benefited from instruction evidenced a decline in recall as assessed during a one week posttest (cf., Keeney et al., 1966; Kennedy & Miller, 1976). These findings suggested that children could be taught to use a verbal strategy which mediated their recall performance but the effects of training were temporary and depended on continued prompting to use what was learned.

The results of the work reviewed above were elaborated further by Hagen, Meacham, and Mesibov (1970) who presented a more difficult version of the serial recall task to adolescents and college students. The data indicated that labeling led to a decrement in the recall of items from the first six of eight serial positions, thus supporting the Hagen and Kingsley (1978) result for their oldest age group, but enhanced memory for recency items. Hagen et al. (1981) summarized this research and stated:

> It is apparent, then, as individuals become increasingly proficient in using their own strategies to facilitate recall, the detrimental effect of an externally imposed strategy, that is not as effective as the subjects' perferred strategy, increases. (p. 31)

The evidence on the development of mediated behavior in children gathered through the use of the selective attention task not only supports the general conclusions cited above but also elaborates our knowledge regarding the purposeful use of verbal behavior. That is, the child must first select the appropriate information to be rehearsed before using a verbal mnemonic (Druker & Hagen, 1969; Sabo & Hagen, 1973). After reviewing much of the research using the central-incidental task, Hagen and Kail (1975) concluded that during the middle school years,

> There appears to be (a) ... tendency to employ task-appropriate strategies ... that certainly involve perceptual processing but also involves a central, cognitive component. (p. 188)

Now consider developmental studies of strategy use among children with cognitive deficiencies. This work deals with the trainability of effective mnemonics. In summarizing many of the training efforts, it appears that the relatively poor memory performance of mentally retarded and learning disabled children results from a lack of efficient (rehearsal) strategy use (cf. Belmont & Butterfield, 1969, 1971; Brown, 1974; Ellis, 1970).

In one study reported by Hagen, Streeter, and Raker (1974), mentally retarded children were trained to use labeling given a serial recall task. Compared to non-handicapped children of equal mental age (mean MA = 7.9 years), the trained mentally retarded youngsters showed a similar recall pattern, i.e., recency performance improved while primacy performance decreased. In a followup study, the retarded youngsters were induced to use a cumulative rehearsal strategy which improved their recall, especially at the primacy positions.

The selective attention task also has been used to assess the cognitive functioning of mentally retarded children and revealed that recall of central information increased with MA whereas memory for incidental material remained relatively constant. Note, however, that the performance of non-handicapped children of comparable chronological age (CA) was superior to that found for retarded children (e.g., Hagen & Huntsman, 1971).

In a related training study, Hagen and West (1970) investigated the effects of instruction on mildly retarded eleven- and fourteen-year-olds' (MA 8 and 10.6 years) recall of primary and secondary aspects of the stimuli used in the selective attention task. These youngsters were instructed to learn both aspects of the pictures and a monetary reinforcement system was used to reward correct recall. However, the payoff for recall of the primary aspects of the stimuli was five times that for recall of secondary aspects. Children at both age levels recalled more primary than secondary information. Thus, these mentally retarded children benefited from this type of instructional procedure.

The conclusion drawn from these studies is that mentally retarded youngsters are deficient with use of effective mnemonic routines but can be trained easily in both attentional and verbal skills. The evidence supports the argument that assumed structural deficits do not account wholly for the poor performance found among mentally retarded children. Instead, through appropriate instruction, attention and memory improve. This view is consistent with the position that mentally retarded children are deficient in the use of certain control processes.

In a more recent study of learning disabled children, Newman and Hagen (1981) investigated the effects of instruction on both serial and free recall. The children ranged in age from seven to thirteen years and were divided into a young and old age group. Even though neither group spontaneously produced a verbal rehearsal strategy, the serial recall of the older group was somewhat higher than that found for the younger children. On the free recall task, in which categorizable items were used, neither group produced an effective organizational routine. However, with instructions the older children's clustering and recall improved. Unfortunately, the performance of the younger children was not affected by the training. This finding points to a number of possible interpretations. It may be that the instructions were not powerful enough to elicit an affective mnemonic or these younger children were in fact mediationally deficient. Regardless of the theoretical explanation, behaviorally these children appeared to be "inactive learners" (Torgesen, 1975).

In other related work, Brown and Barclay (1976) and Brown et
al. (1979) reported on the maintenance and generalization of trained
mnemonics among mentally retarded youngsters (also see Campione &
Brown, 1977). In the initial study (Brown & Barclay, 1976), independent groups of mentally retarded children (MA 84 and 105 months;
CA 116 and 144 months) were trained to use different study strategies given a serial (ordered) recall task. The children's actual
memory spans were assessed and supra-span sets of pictures, i.e.,
span + 1/2, were used in all subsequent testing. For example, if a
child's actual span equalled 6-items, then the supra-span sets contained 9-items.

The study was designed such that MA (2 levels: 6 and 8 years),
type of training (3 levels: labeling, rehearsal, integration), and
posttest delay (3 levels: immediate prompted, 1-day delayed unprompted, 2-week delayed unprompted) were factorially combined.

The children were required to self-present to-be-remembered
pictures one at a time in sequence and allowed to study the item
until they felt ready for a recall test (cf. Flavell et al., 1970).
In a pretest condition it was found that all children performed
poorly. No significant recall differences were revealed among
training groups within each MA level. In subsequent sessions the
children were trained to use one of three study strategies: labeling,
rehearsal, or anticipation.

The children in the labeling group were instructed to look at
and name each item. This procedure was repeated four times for each
of three picture sets. Youngsters receiving rehearsal instructions
were told to name each picture on the first pass through the list;
then, for three additional exposures they were trained to rehearse
cumulatively, e.g., expose and name three consecutive pictures and
repeat the names as a set three times over. In the anticipation
training group, the children began by familiarizing themselves with
the pictures through naming. This was followed by a number of
trials in which the children were requested to generate the name of
the picture before looking at it. The rehearsal and anticipation
training conditions were designed so that the children self-tested
themselves during the study period. Each child was trained on three
lists a day for two days. Following training, the effects of instruction were assessed in three posttest sessions.

An immediate posttest was given the day after the completion
of training. Here, the children were reminded (prompted) that they
had been instructed in the use of a certain strategy. The next two
posttests were unprompted, one given the day following the prompted
posttest and the other two weeks later.

The interesting findings were that for each MA level, rehearsal and anticipation training significantly improved recall on the prompted posttest relative to the label group. However, on the un-prompted posttests the younger (MA 6) children abandoned the trained strategies which resulted in poorer recall; a result which repli-cated that reported by Keeney et al. (1967), Kennedy and Miller (1976), and others. The performance of the MA 8 children was more encouraging since they continued to use the trained routines without prompting. Apparently, the older children trained to rehearse or anticipate recognized the relationship between their strategy efforts and improved performance. That is, the procedures used here provided the mnemonics to meet the task demands as well as the knowledge required to self-maintain a learned strategy.

In a followup study, Brown et al. (1979) retested all the children trained in the previous experiment after a 1-year delay. This research extended the Brown and Barclay (1976) findings by presenting a "far generalization" test of the effects of instruc-tion -- a prose recall task was used. Using a procedure similar to that in the initial study, each youngster was retested on a serial recall task during four posttest sessions -- one session per day for four consecutive days -- the first, second, and fourth posttests were unprompted; on day three all children were prompted to use the mnemonic routine they had already learned.

The data from the first two unprompted posttests showed that the MA 6 children did not maintain the trained strategies while the older children showed clear evidence of maintenance. On the prompted posttest the younger children's recall improved signifi-cantly as did that of the MA 8 youngsters; yet on the next un-prompted test the younger children abandoned the trained strategies. These results point again to the transitory effects of instruction on the unprompted maintenance of young children. Perhaps explicit explanations of metamemorial variables or extended training would result in a more promising outcome. In contrast, the older chil-dren benefited from the self-testing component of rehearsal and an-ticipation training which lead to the self-control of strategy use.

Consider next the results from presenting a generalization prose recall task to the children at MA 8 years. The younger chil-dren were not tested here since they showed no evidence of main-tenance on a task similar to the one on which they were trained. Also, a new sample of naive older youngsters were selected as a no training control. Each child was tested during six sessions which occurred on different days. In any one session, the children read two stories with instructions to study the material until they could tell the story back to the experimenter in detail. The stories had been rated previously by adults who made judgments regarding the number of the idea units in each passage and the importance level of each unit to the comprehension of the total story (cf. Johnson, 1970).

Analyses of the prose recall data indicated that the children in the rehearsal and anticipation training groups remembered significantly more idea units than the youngsters in the label and no training contrast groups. Further, the training factor interacted with importance level; children instructed to rehearse or anticipate recalled reliably more information from the highest level than did the control children. The results of this experiment are encouraging in at least two ways. First, the effects of instruction were evident a year after training, and; second, the self-monitoring component of the rehearsal and anticipation training procedures generalized to a new situation. It seems reasonable to conclude that one set of optimal conditions for inducing generalization includes task demands which sharpen the children's metacognitive skills -- especially memory monitoring of the person, task, and strategy relationships (see also, Barclay, 1979; Brown, 1978; Butterfield, Siladi, & Belmont, 1980; Kestner & Borkowski, 1979; Newell & Barclay, 1981, for a similar conclusion).

The research reviewed in this section suggested that the necessary and sufficient conditions for the unprompted maintenance and generalization of strategic behavior include the following: a) that children produce and know how to use a mnemonic device; and, more importantly, b) that the youngsters become aware of the causative relationship between strategy use and improved performance -- instructional procedures must instantiate the memory-metamemory connection. Note further that neither condition is a dichotomy in the sense that the child either produces a strategy and uses it knowledgeably or not; instead, the acquisition of strategies and metamemorial awareness represent two interrelated continua (Russell, 1913).

In the next section these conclusions afford the conceptual frame of reference through which intervention programs designed to remediate learning disabilities are based. In fact, Torgesen (1977a) provides some of the rationale supporting the view that many learning disabled children are both strategy and knowledge deficient with respect to certain problem solving behaviors. He states,

... poor performance in many different task settings may be due to the child's failure to actively engage the task through the use of efficient strategies and other techniques of intelligence.

Further study of the role of certain 'meta' variables in the learning performance of learning disabled children seems essential for two reasons. First, such study would necessarily lead to a more careful examination of the construct validity of measures which have been used to study the specific psychological defects of learning

disabled children ... Second, greater understanding of
the ways in which these nonspecific (metacognitive)
factors influence performance in a variety of situations
would provide information useful for remedial purposes.
(p. 39)

IV.   QUESTIONS AND ISSUES FOR FUTURE RESEARCH IN LEARNING DISABILITIES

This chapter is organized to present the position that a learning disability may be portrayed in terms of the child's skills and
deficiencies in the use of certain problem solving strategies. Most
efforts aimed at defining learning disabilities to date have been
concerned with identifying who is disabled and how the individuals
so labeled are impaired. Because of the tremendous heterogeniety of
the population studied, there has been considerable disagreement in
the application of the label. In practice, then, children who do
not achieve well overall in school get labeled as learning disabled.
A question which arises is whether a learning disability is more a
social phenomenon than a psychological reality.

The position being taken here is based on the assumption that
a learning disability is not an entity or clearly marked dysfunction
residing solely in the child. Rather, the child is an integral part
of his or her learning environment. When either neurophysiological
or environmental factors are considered in isolation, then the
interrelations between them may be ignored. Future research must be
aimed at explicating the "fit" between the maturing child and the
demands of the educational system. The child's performance, then,
is viewed relative to a developmental, social, and cultural context.

A counter position to this perspective focuses on the structural integrity of the child's physiological system. Both central and
peripheral aspects of the system must be intact for information
processing to proceed normally. The young child is not "ready" to
engage in adult-like processing, while the older, learning disabled
child has not developed in the same way as have his or her peers.

Any integrated model of learning disabilities surely would include a consideration of neurological features. However, it is not
reasonable to assume that such features account totally for the
difficulties learning disabled children have in acquiring knowledge
through reading, mathematics, or other school-related tasks. The
child's intentions, beliefs, and motivation to learn (cf. Dweck,
1981; Paris, 1978), together with acquired strategies, knowledge,
and the demands of schooling must all be taken into account if one
purports to offer a complete explanation of the problem.

Hagen et al. (1981) have suggested an integrated view of learning disabilities. That is, a learning disability can be seen as a "fuzzy concept" (cf. Rosche, 1975, 1978) defined in terms of behavioral response patterns over various types of tasks, e.g., reading or memory. Two factors make this model appealing to us. First, the fuzzy concept notion acknowledges the current state of affairs concerning the definition of learning disabilities. Second, it allows for, indeed recommends, a functional characterization of what a learning disability is. The concept of "learning disability" can then easily be defined to include the notions of the child's assessed strategy usage and knowledge base.

This model (Hagen et al., 1981) includes two major variables: an analysis of distinctive characteristics and demands of specific tasks, and an evaluation of what the child does (i.e. strategies) and knows given certain tasks. This second variable also includes information regarding the child's cognitive style or activity (Torgesen, 1975, 1977a) during a problem solution. It may be that by the time children are identified as having major difficulties learning in academic situations, they are convinced that they cannot deal with the tasks appropriately and hence do not attempt to apply their skills and knowledge.

Consider the case of reading comprehension. Reading requires a flexible set of problem solving routines. The more general strategies, such as scanning the lines, comprehension checking, and self-monitoring, must be flexibly employed to meet the demands of different types of material as well as different purposes for reading. Further, knowledge of reading as an activity together with prior knowledge of the subject matter facilitate comprehension. Through an analysis of reading tasks, it may be possible to identify the various perceptual and cognitive skills needed to understand written material. If this approach is employed successfully, it should provide not only an understanding of the child's actual performance in a reading comprehension task, but it should also lead directly to appropriate intervention measures.

The potentially useful consequence of this approach, then, is that a priori patterns of task performance can be derived (cf. Siegler, 1976, 1978, 1980; Sternberg, 1979). These provide a frame of reference for evaluating the cognitive skills and related functioning of a given child. The degree to which the child's actual behavior deviates from the expected affords a basis for instruction.

The perspective offered in this chapter is that potentially useful research on learning disabilities should focus on how children use appropriate strategies and existing knowledge to acquire new information in a purposeful manner. An important thrust of this research would be towards explaining the relationship between the child's current cognitive-developmental status, the demands of schooling, and the context in which he or she is expected to function. From the developmental perspective, three questions would be addressed by such research: (a) the transition from being a novice or ineffective learner to becoming an "expert" problem solver; (b) access of existing knowledge as needed to meet the demands of new or novel problems; and (c) the fit or compatibility between what is already known and the demands of a new task (cf. Chi & Brown, 1981). These questions, in turn, would have to be considered in light of children's belief systems about their own cognitive capabilities and limitations when asked to learn in school.

References

Asarnow, J. R., & Meichenbaum, D.   Verbal rehearsal and serial recall: The motivational training of kindergarten children. Child Development, 1979, 50, 1173-1177.

Atkinson, R. C., & Shiffrin, R. M.   Human memory: A proposed system and its control processes. In K. W. Spence and J. T. Spence (Eds.), The psychology of learning and motivation (Vol. 2). New York: Academic Press, 1968.

Barclay, C. R.   A component view of memory development. Psychology: A study of human behavior, 1981, in press.

Barclay, C. R.   The executive control of mnemonic activity. Journal of Experimental Child Psychology, 1979, 27, 262-276.

Bauer, R. H.   Memory, acquisition, and category clustering in learning disabled children. Journal of Experimental Child Psychology, 1979, 27, 365-383.

Bauer, R. H.   Memory processes in children with learning disabilities: Evidence for deficient rehearsal. Journal of Experimental Child Psychology, 1977, 24, 415-430.

Bebko, J. M.   Can recall differences among children be attributed to rehearsal effects? Canadian Journal of Psychology, 1979, 33, 96-105.

Belmont, J. M., & Butterfield, E. C.   What the development of short-term memory is. Human Development, 1971, 14, 236-248.

Belmont, J. M., & Butterfield, E. C.   The relations of short-term memory to development and intelligence. In L. Lipsitt and H. Reese (Eds.), Advances in child development and behavior (Vol. 4). New York: Academic Press, 1969.

Brown, A. L.   Knowing when, where, and how to remember: A problem of metacognition. In R. Glaser (Ed.), Advances in instructional psychology. Hillsdale, N.J.: Erlbaum, 1978.

Brown, A. L.   The development of memory: Knowing, knowing about knowing, and knowing how to know. In H. W. Reese (Ed.), Advances in child development and behavior (Vol. 10). New York: Academic Press, 1975.

Brown, A. L.   The role of strategic behavior in retardate memory.
    In N. R. Ellis (Ed.), International review of research in mental
    retardation (Vol. 7). New York: Academic Press, 1974.
Brown, A. L.   Judgements of recency for long sequences of pictures:
    The absence of a developmental trend. Journal of Experimental
    Child Psychology, 1973, 15, 473–481.
Brown, A. L., & Barclay, C. R.   The effects of training specific
    mnemonics on the metamnemonic efficiency of retarded children.
    Child Development, 1976, 47, 70–80.
Brown, A. L., & Campione, J. C.   The effects of knowledge and
    experience on the formation of retrieval plans for studying from
    text. In M. M. Greenberg, P. E. Morris, and R. N. Sykes (Eds.),
    Practical aspects of memory. New York: Academic Press, 1978.
Brown, A. L., Campione, J. C., & Barclay, C. R.   Training self-
    checking routines for estimating recall readiness: Generalization
    from list learning to prose recall. Child Development, 1979, 50,
    501–512.
Brown, A. L., Campione, J. C., & Murphy, M. D.   Maintenance and
    generalization of trained metamnemonic awareness by educable
    retarded children: Span estimation. Unpublished manuscript,
    University of Illinois, 1976.
Brown, A. L., & Scott, M. S.   Recognition memory for pictures in
    preschool children. Journal of Experimental Child Psychology,
    1971, 11, 401–412.
Butterfield, E. C., Siladi, D., & Belmont, J. M.   Validating
    theories of intelligence. In H. Reese and L. Lipsitt (Eds.),
    Advances in child development and behavior (Vol. 15). New York:
    Academic Press, 1980.
Buyer, D., & Hagen, J. W.   Selective attention and memory in low
    achieving, neurologically impaired and normal children. Paper
    presented at the biennial meetings of the Society for Research
    in Child Development, Boston, April, 1981.
Campione, J. C., & Brown, A. L.   Memory and metamemory development
    in educable retarded children. In R. V. Kail and J. W. Hagen
    (Eds.), Perspectives on the development of memory and cognition.
    Hillsdale, N.J.: Erlbaum, 1977.
Chi, M. T. H.   Knowledge structures and memory development. In
    R. Siegler (Ed.), Children's thinking: What develops? Hills-
    dale, N.J.: Erlbaum, 1978.
Chi, M. T. H., & Brown, A. L.   The development of knowledge and
    expertise. Paper presented at the biennial meetings of the
    Society for Research in Child Development, Boston, April, 1981.
Cronback, L. J.   The two disciplines of scientific psychology.
    American Psychologist, 1957, 12, 671–684.
Dempster, F. N.   Short-term storage capacity and chunking. A
    developmental study. Unpublished Doctoral Dissertation, Univer-
    sity of California, 1977.
Druker, J. F., & Hagen, J. W.   Developmental trends in the pro-
    cessing of task-relevant and task-irrelevant information. Child
    Development, 1969, 40, 371–382.

Dweck, C.   Theories of intelligence and achievement motivation.
    Paper presented at the Summer Institute on Learning and Motiva-
    tion in the Classroom. University of Michigan, 1981.
Ellis, N. R.   Memory process in retardates and normals. In N. R.
    Ellis (Ed.), International review of research in mental retar-
    dation (Vol. 4). New York: Academic Press, 1970.
Flavell, J. H.   Developmental studies of mediated memory. In H. W.
    Reese and L. P. Lipsitt (Eds.), Advances in child development
    and behavior (Vol. 5). New York: Academic Press, 1970.
Flavell, J. H., Beach, D. H., & Chinsky, J. M.   Spontaneous verbal
    rehearsal in a memory task as a function of age. Child Develop-
    ment, 1966, 37, 283-299.
Flavell, J. H., Friedrichs, A. G., & Hoyt, J. D.   Developmental
    changes in memorization processes. Cognitive Psychology, 1970,
    1, 324-340.
Flavell, J. H., & Wellman, H. M.   Metamemory. In R. V. Kail and
    J. W. Hagen (Eds.), Perspectives on the development of memory
    and cognition. Hillsdale, N.J.: Erlbaum, 1977.
Gelabert, T., Torgesen, J., Dice, C., & Murphy, H.   The effects of
    situational variables on the use of rehearsal by first-grade
    children. Child Development, 1980, 51, 902-905.
Gibson, E. J.   Principles of perceptual learning and development.
    New York: Appleton-Century-Crofts, 1969.
Greeno, J.   Meaningful learning in geometry and algebra. Paper
    presented at the Summer Institute on Learning and Motivation in
    the Classroom. University of Michigan, 1981.
Hagen, J. W.   Strategies for remembering. In S. Farnham-Diggory
    (Ed.), Information processing in children. New York: Academic
    Press, 1972.
Hagen, J. W.   Some thoughts on how children learn to remember.
    Human Development, 1971, 14, 262-271.
Hagen, J. W.   The effect of distraction on selective attention.
    Child Development, 1967, 38, 685-694.
Hagen, J. W., Barclay, C. R., & Schwethelm, B.   Cognitive develop-
    ment of the learning disabled child. In N. Ellis (Ed.) Inter-
    national review of research in mental retardation. New York:
    Academic Press, 1981.
Hagen, J. W., & Hale, G. A.   The development of attention in chil-
    dren. In A. Pick (Ed.), Minnesota symposia on child psychology
    (Vol. 7). Minneapolis: University of Minnesota Press, 1973,
    117-140.
Hagen, J. W., Hargrave, S., & Ross, W.   Prompting and rehearsal in
    short-term memory. Child Development, 1973, 44, 201-204.
Hagen, J. W., & Huntsman, N.   Selective attention in mental retar-
    dates. Developmental Psychology, 1971, 5, 151-160.
Hagen, J. W., Jongeward, R. H. Jr., & Kail, R. V. Jr.   Cognitive
    perspectives on the development of memory. In H. Reese (Ed.),
    Advances in child development and behavior (Vol. 10). New York:
    Academic Press, 1975.

Hagen, J. W., & Kail, R. V. The role of attention in perceptual and cognitive development. In W. Cruickshank and D. Hallahan (Eds.), Perceptual and learning disabilities in children (Volume 2: Research and theory. New York: Syracuse University Press, 1975.

Hagen, J. W., & Kingsley, P. R. Labeling effects in short-term memory. Child Development, 1968, 39, 113-121.

Hagen, J. W., Meacham, J. A., & Mesibov, G. Verbal labeling, rehearsal, and short-term memory. Cognitive Psychology, 1970, 1, 47-58.

Hagen, J. W., & Stanovich, K. E. Memory: Strategies of acquisition. In R. V. Kail Jr. and J. W. Hagen (Eds.), Perspectives on the development of memory and cognition. Hillsdale, N.J.: Erlbaum, 1977.

Hagen, J. W., Streeter, L. A., & Raker, R. Labeling, rehearsal, and short-term memory in retarded children. Journal of Experimental Child Psychology, 1974, 18, 259-268.

Hagen, J. W., & West, R. The effects of a payoff matrix on selective attention. Human Development, 1970, 13, 43-52.

Hallahan, D. P., & Cruickshank, W. M. Psychoeducational foundations of learning disabilities. Englewood Cliffs, N.J.: Prentice Hall, 1973.

Johnson, R. E. Recall of prose as a function of the structural importance of linguistic units. Journal of Verbal Learning and Verbal Behavior, 1970, 9, 12-20.

Kahneman, D. Attention and effort. Englewood Cliffs, N.J.: Prentice Hall, 1973.

Keeney, T. J., Cannizzo, S. R., & Flavell, J. H. Spontaneous and induced verbal rehearsal in a recall task. Child Development, 1967, 38, 953-966.

Kendler, H. H., & Kendler, T. S. Vertical and horizontal processes in problem solving. Psychological Review, 1962, 69, 1-61.

Kennedy, B. A., & Miller, D. J. Persistent use of verbal rehearsal as a function of information about its value. Child Development, 1976, 47, 566-569.

Kestner, J., & Borkowski, J. C. Children's maintenance and generalization of an interrogative strategy. Child Development, 1979, 50, 485-494.

Kingsley, P. R., & Hagen, J. W. Induced versus spontaneous rehearsal in short-term memory in nursery school children. Developmental Psychology, 1969, 1, 40-46.

Kobasagawa, A. Retrieval strategies in the development of memory. In R. V. Kail and J. W. Hagen (Eds.), Perspectives on the development of memory and cognition. Hillsdale, N.J.: Erlbaum, 1977.

Kreutzer, M. A., Leonard, C., & Flavell, J. H. An interview study of children's knowledge about memory. Monographs of the Society for Research in Child Development, 1975, 40, (1, Serial No. 159).

Landau, B., & Hagen, J. W.   The effect of verbal cues on concept
    acquisition and retention in normal and educable mentally re-
    tarded children. Child Development, 1974, 45, 643-650.
Luria, A. R.   Cognitive development: Its cultural and social
    foundations. Cambridge, MA: Harvard University Press, 1976.
Maccoby, E. E., & Hagen, J. W.   Effects of distraction upon central
    versus incidental recall: Developmental trends. Journal of
    Experimental Child Psychology, 1965, 2, 280-289. (Also appears
    in Adolescent Development, J. Hill and J. Shelton (Eds.),
    Englewood Cliffs, N.J.: Prentice-Hall, 1971, 113-120).
Mandler, J. M., & Robinson, C. A.   Developmental changes in picture
    recognition. Journal of Experimental Child Psychology, 1978,
    26, 122-126.
Meyers, N. A., & Perlmutter, M.   Memory in the years from two to
    five. In P. A. Ornstein (Ed.), Memory development in children.
    Hillsdale, N.J.: Erlbaum, 1978.
Miller, G. A.   The magical number seven, plus or minus two: Some
    limits on our capacity for processing information. Psychological
    Review, 1956, 63, 81-97.
Newell, K. M., & Barclay, C. R.   The development of knowledge about
    action. In J. A. S. Kelso and J. Clark (Eds.), Development of
    human motor skills. New York: Wiley, 1981, in press.
Newman, R. S., & Hagen, J. W.   Memory strategies in children with
    learning disabilities. Journal of Applied Developmental
    Psychology, 1981, 1, 297-312.
Paris, S. G.   Coordination of means and goals in the development
    of mnemonic skills. In P. A. Ornstein (Ed.), Memory development
    in children. Hillsdale, N.J.: Erlbaum, 1978.
Piaget, J.   The child and reality: Problems of genetic psychology.
    New York: Penguin Books, 1976.
Piaget, J.   Biology and knowledge. Chicago: University of Chicago
    Press, 1971.
Piaget, J.   The origins of intelligence in children. New York:
    International University Press, 1952.
Reese, H. W.   Verbal mediation as a function of age level.
    Psychological Bulletin, 1962, 59, 502-509.
Resnick, L. B.   Toward a cognitive theory of instruction. Paper
    presented at the Summer Institute on Learning and Motivation in
    the Classroom. University of Michigan, 1981.
Rosche, E. R.   Human categorization. In N. Warren (Ed.), Studies
    in cross-cultural psychology. London: Academic Press, 1978.
Rosche, E. R.   Universals and specifics in human categorization.
    In R. Brislin, S. Bochner, and W. Lonnen (Eds.), Cross-cultural
    perspectives on learning. New York: Halsted, 1975.
Russell, B.   On the notion of cause. Proceedings of the Aristo-
    telian Society (New Series), 1913, 13, 1-26.
Sabo, R. A., & Hagen, J. W.   Color cues and rehearsal in short-
    term memory. Child Development, 1973, 44, 77-82.

Siegler, R. S. Recent trends in the study of cognitive development: Variations on a task-analytic theme. Human Development,
1980, 23, 278-285.

Siegler, R. S. The origins of scientific reasoning. In R. S.
Siegler (Ed.), Children's thinking: What develops? Hillsdale,
N.J.: Erlbaum, 1978.

Siegler, R. S. Three aspects of cognitive development. Cognitive
Psychology, 1976, 4, 481-520.

Simon, H. A. How big is a chunk? Science, 1974, 183, 482-488.

Smiley, S. S., & Brown, A. L. Conceptual preference for thematic
or taxonomic relations: A non-monotonic age trend from preschool
to old age. Journal of Experimental Child Psychology, 1979, 28,
249-257.

Smiley, S. S., & Weir, M. W. The role of dimensional dominance in
reversal and nonreversal shift behavior. Journal of Experimental
Child Psychology, 1966, 4, 296-307.

Sternberg, R. J. The nature of human abilities. American
Psychologist, 1979, 34, 214-230.

Torgesen, J. K. Conceptual and educational implications of the use
of efficient task strategies by learning disabled children.
Journal of Learning Disabilities, 1981, in press.

Torgesen, J. K. The role of non-specific factors in the task performance of learning disabled children: A theoretical assessment. Journal of Learning Disabilities, 1977a, 10, 33-41.

Torgesen, J. K. Memorization processes in reading-disabled children. Journal of Educational Psychology, 1977b, 69, 571-578.

Torgesen, J. K. Problems and prospects in the study of learning
disabilities. In E. M. Hetherington (Ed.), The review of child
development research (Vol. 5). Chicago: The University of
Chicago Press, 1975.

Torgesen, J. K., & Goldman, T. Verbal rehearsal and short-term
memory in reading disabled children. Child Development, 1977,
48, 56-60.

Vygotsky, L. S. Thought and Language. Cambridge: MIT Press, 1962.

Wellman, H. Knowledge of the interaction of memory variables: A
developmental study of metamemory. Developmental Psychology,
1979.

READING DISABILITY AND ITS RELATION TO INFORMATION INTEGRATION

J. P. Das, F. Snart, and R. F. Mulcahy

Centre for the Study of Mental Retardation and
Department of Educational Psychology
University of Alberta, Edmonton, Canada

INTRODUCTION

The recent upsurge of publications on reading disability, variously named as dyslexia and minimal brain dysfunction, has generated significant issues on which agreement is lacking. Perhaps the most frequently raised issue is the identification of the population. It is sometimes clinically possible to diagnose a child as a dyslexic rather than a poor reader. Such children are typically bright, and even articulate. They can tell a story which would be better than the average story written by their agemates; but they have minimal competence in reading single words, and enormous problems in reading sentences. Thus, they can be distinguished from the poor reader who is by no means bright, and often has a performance IQ between 85 and 90. He seems to be poor in almost all cognitive tasks including language comprehension and speech production. But as Frith (1981) observes, it is not always easy to identify a group of dyslexics, especially when conventional educational tests are applied. We are not going to discuss the issue of diagnosis in this chapter; the problem has been raised and discussed in several chapters in this book. Instead, we shall turn to another important problem in the field of learning disability, which is the identification of basic cognitive processes underlying reading backwardness. We shall also present a discussion of methods to remediate observed deficits in cognitive processes.

In reviewing the recent literature on cognitive processes and reading, one is impressed with certain outstanding conclusions. Although these conclusions are limited by our knowledge of reading and its difficulties at the present time, nevertheless they seem to

85

view reading difficulties as difficulties in information processing.
Four such conclusions regarding reading difficulties are described
below.

1. The disabled reader is characterized essentially by a verbal
deficit. The verbal deficit most probably relates to verbal encoding
as well as to other linguistic processes which are involved in understanding syntax and meaning. Short-term memory problems by themselves do not characterize the reading disabled individual; rather,
the difficulty appears to be in the rapid coding of linguistic
information in short-term memory. Vellutino (1977) and Perfetti and
Goldman (1976) have contributed to this view. However, this view has
been challenged by research on eye movement; the dyslexics appear to
engage in regressive eye movements (Pavlidis, 1981).

2. The disabled reader suffers from a defective control
process in terms of memory. Some reading disabled individuals do not
use the right sort of mnemonic codes and hence cannot store and
recall information at a later time. Torgesen (1980, and chapter in
this book) has done the most to draw our attention to this problem.

3. The disabled reader has a deficit in phonological coding.
Whereas the good reader uses both visual or graphemic and phonological strategies in reading, the poor reader has an inadequate and
inconsistent phonological coding strategy. It seems that the poor
reader uses visual-orthographic strategies for reading and a phonological strategy for spelling. Since information cannot be stored
and retrieved easily if it has been coded by a visual orthographic
strategy, the poor reader cannot use such information in understanding what he has read (Bradley & Bryant, 1978). Lesgold and
Perfetti's (1981) view of reading disability overlaps, to a certain
extent with this view, although they would emphasize slowness in
phonological coding as the real reason for poor comprehension.

The role of phonological coding requires some clarification.
Phonological coding can operate in one of two ways according to
Underwood (1978). Coding is done through grapheme-phoneme translation (e.g., school, the grapheme "sch" corresponds to a phoneme),
and through grapheme-syllabic correspondence (e.g., wool). The
fluent reader who also comprehends well takes advantage of phonological coding because through it, (a) memory of the word is preserved
while reading a sentence, (b) the written material can be given
intonations and emphases not found in print, and (c) phonological
recoding converts print, which is visual-spatial, to an articula-
tory-sequential mode. It is precisely this demand of conversion to
a sequential order which the poor reader cannot meet. The purpose
of both kinds of coding is to store the words in memory in order to
comprehend a sentence or any connected discourse.

Words can be recognized and single words can be understood through direct visual access without phonological coding. Good readers use both, when appropriate, whereas poor readers seem to be using a visual-orthographic strategy for reading. Perhaps it is because of their difficulty in generating phonological codes that they do not give up using orthographic coding (Barron, 1980). Good readers are penalized for their zealous use of phonological coding. They are more confused by pseudohomophones (e.g., baul) which sound like, but are not spelled like a word (Barron, 1980), and they lose their advantage of superior memory in recalling strings of confusable consonants (e.g., BCPTD) because of phonological coding (Shankweiler, Liberman, Mark, Fowler, & Fischer, 1979).

4. The disabled reader has problems in organization of strategic behavior. Reading and comprehension require organization strategies, in short, plans. These are goal-directed activities. Bradley and Bryant (1978), Underwood (1978), and Leong (1974) have described the important role that strategies and organization play in reading. What is required beyond analysis, discrimination and encoding is essentially forming hypotheses about the text and testing these hypotheses. Underwood (1978) succinctly summarizes this by saying:

> ... that the text might be described as an approximation of what the writer wishes to convey. The reader may hold a series of approximate meanings of the text before the comprehension of the meaning as intended by the writer.

Each of the four problems of the disabled reader refers to a deficit in information processing, to certain difficulties in acquisition, analysis, storage, and selection of information for the purpose of fulfilling a goal. They do not refer to deficits in visual processing or to other sensory and perceptual handicaps. In short, reading disability is not attributed to defects in sensory modalities.

INFORMATION-INTEGRATION MODEL AND ITS APPLICATION

A model of information processing, or rather information-integration, has been developed following the notions of Luria (1966). It will be described here briefly, and then some results with applications to reading disability will be presented. Finally, certain methods of remediation of reading disability will be discussed.

Simultaneous and successive processing are two modes of coding or information-integration originally suggested by Luria (1966) on the basis of his observations of different types of cortical lesions

and their behavioral correlates. Simultaneous processing involves
the formation of a code that is quasi-spatial in nature, having the
characteristic that all parts of it are immediately surveyable; the
code can operate on verbal as well as nonverbal information. Successive processing involves the formation of a code that is more
temporal in nature, being accessible only in a linear way. Like
simultaneous processing, successive coding can be applied to verbal
and nonverbal information as well as to information from any modality such as visual and auditory. According to Luria, simultaneous
processing is linked to the occipito-parietal areas of the cortex,
and successive processing to the fronto-temporal areas. Factors
identifiable as simultaneous and successive processing have emerged
in a large number of studies, and we have attempted to explore their
relationships to traditional models of abilities, intelligence, and
language (see Das, Kirby, & Jarman, 1979).

Planning is broadly defined as the generation, selection, and
execution of programs, and is located in the frontal lobes, especially in the prefrontal area (Das, 1980). Planful behavior entails
searching for the most important aspect of information present,
comparing each part of information with others parts, creating
hypotheses and the verification of hypotheses. Along with the two
coding processes, simultaneous and successive, a planning factor,
which is orthogonal to the coding factors has been obtained in a
variety of samples of adults, children, and mentally retarded individuals (Das, 1980). Planning as a process should be observed in
all individuals after a certain expertise in coding has developed.

We shall first report an experiment on reading disability
carried out by Das and Snart in order to illustrate the application
of the information-integration model. This will be followed by
summaries of some further studies.

An illustrative study

As mentioned in the introduction, memory functions have been
implicated in reading disability. Sometimes, the test of memory is
Digit Span (Torgesen & Houck, 1980), and at other times, an exhaustive battery of memory tasks has been used (Nelson & Warrington,
1980) in order to distinguish the reading disabled from normal or
average readers. We have also noted in the introduction, that a
related, but somewhat separate deficit in verbal processes has been
attributed to disabled readers. Vellutino (1977) has supported such
a view in his well-known paper. Since reading essentially involves
verbal encoding, and a prerequisite for comprehension is to hold in
memory what has recently been read, deficits in memory and verbal
skills are implicated in reading disability. However, in tasks
which require little memory or verbal skills, the dyslexics are
still found to be inferior to normal readers as reported by Das,
Leong, and Williams (1978).

Leong (1974) tested 58 boys (9- to 10-years-old), who have average nonverbal IQ, but were reading at least 2.5 grades behind their peers of the same age. A comparable group of 58 average readers were used as controls. The object of testing was to determine if the disabled readers were less competent in successive coding tasks (Visual Short-term Memory and Auditory Serial Recall of words). As suggested in the literature, they are expected to be poor in sequencing and generally in short-term memory. In contrast to the successive tasks, the simultaneous coding tasks were Raven's Coloured Progressive Matrices, Figure Copying and Memory for Designs. In these, the disabled readers should not be inferior to the controls. However, their performance was found to be poor in all tasks. It was then suggested that the difficulty lies in adopting appropriate strategies when novel tasks are presented to the disabled reader, that they are particularly deficient in planning or strategic behavior. Using Digit Span tests, Torgesen and his associates (cf. Torgesen & Houck, 1980) have similarly concluded that reading disabled children will have difficulties in any task that requires planful organization of information.

Some tasks which have loaded on a 'planning' factor in previous studies (Das, 1980) were therefore selected for administration to disabled readers. These children were also given Figure Copying and a test for Serial Recall of words as in Leong's (1974) experiment in order to reconfirm, if possible, his results with the samples of disabled readers in the present study. A major objective of the study, was to contrast average and disabled readers across two age groups on the above tasks. It will be clear in the Method section that dividing subjects by age will enable us to determine the extent to which the performance of the disabled reader fell behind the performance of average readers.

Thus, a major objective of the study was to contrast the reading disabled and control children across two age groups on planning tasks as well as on coding; the latter would be an attempt to verify coding deficits already observed by Leong (1974). The second objective of this study was to determine the extent to which the performance of the disabled reader fell behind the performance of controls. The disabled reader was at least 2 years behind in reading in this study, as in most other studies. Was his performance in coding and planning tasks also retarded by 2 years?

The disabled reader, when compared to the average reader, may show both qualitative and quantitative differences in cognitive processing; in some tasks discontinuities may be noticed, in others, his performance may appear to be merely delayed. An interesting research question to ask is this: Among the average readers, do the relatively delayed readers show the same pattern of incompetencies in a set of cognitive tasks as the disabled readers? By dividing

average readers into those who are below versus those who are above
the median in reading proficiency, we have attempted to answer the
question.

Forty-five average readers from Grade 4 and 45 from Grade 6
were selected for comparison with 15 disabled readers matched on age
for each grade. The specific selection procedure is as follows:
Fifteen students from regular Grade 4 classes, 15 from regular Grade
6 classes, and 10 students from a special class for children with
learning problems were selected from each of three elementary
schools in Edmonton. The 30 children from special classes thus
selected were two years or more behind their agemates in reading,
based on the Edmonton Elementary Reading Tests of decoding and comprehension. They were divided into sub-groups based upon age, such
as 15 reading disabled children (Grade 6 RD) were of comparable age
to the 45 average readers in Grade 6 and 15 reading disabled children (Grade 4 RD) to the 45 in Grade 4. The mean ages of the four
groups were as follows: 118.5 months for Grade 4, 123.8 months for
Grade 4 RD, 142.0 months for Grade 6, and 142.2 months for Grade 6
RD. In terms of age, $\underline{t}$-tests indicated no difference either between
the Grade 4 and Grade $\overline{4}$ RD, or between Grade 6 and Grade 6 RD groups.

In choosing the subjects from special classes, no attempt was
made to equate the IQ's of this group with those of the regular
class subjects. We did not wish to restrict the sample characteristics in any way except that the special class children were backward
in reading by two years or more, had no overt emotional problems, and
were not mentally retarded as considered by the school board.

Apart from Figure Copying and Auditory Serial Recall tasks as
given by Leong (1974), three planning tasks were used in the present
study. In the choice of tasks, we were guided by the results of
previous factor analytic work (Das, 1980). In addition to these, a
test of 3-term syllogistic reasoning was administered to the subjects. Syllogistic reasoning perhaps requires major cognitive processes such as spatial and temporal processing, and planning in
order to verify if the conclusion follows from the premises; in
short, all of the processes implicated in the other tasks.

Figure Copying was originally developed by the Gesell Institute
(Ilg & Ames, 1964), and has been used as a marker task for simultaneous coding (Das, 1973a, Kirby & Das, 1977). Ten geometrical
figures are presented, one at a time, with a space below each into
which the subject copies the design. Figure Copying does not involve a memory component, as the child merely copies figures while
they remain in view. Each drawing response is scored 0, 1, or 2,
according to the maintenance of geometric relations and proportions.

Auditory Serial Recall is a marker for successive processing. This test consists of 16 lists of words, progressing in length from 4 to 7 words. Each series of words is read to the subject at the rate of approximately one word per second, and the subject repeats as many of the words as are recalled. The score consists of the total number of words recalled in the correct serial order.

The next three tests, Trail Making, Visual Search, and Planned Composition, have been found to load together on a planning factor (Ashman & Das, 1980). The Trail Making test was adopted by Armitage (1946), Reitan (1955), and Spreen and Gaddes (1969) as a neurological screening test for brain damage. In the original test, Part A, subjects connected encircled numbers which were distributed randomly over a page, in the correct sequence, as quickly as possible. In Part B, they were required to connect letters and numbers in increasing sequence such as 1-A-2-B-3-C, and so on. The score was based on total elapsed time for task completion. In the present study, however, two new items, plus a practice item were created based on the original format of the Trail Making task. In one of the two modified forms of the test, numbers from 1 to 16 were arranged in two vertical columns separated by 13 cms; the distance between the vertically adjacent numbers was 2 cms. The numbers in one column were 1, 4, 5, 8, 9, 12, 13, 16, whereas in the other, they were 2, 3, 6, 7, 10, 11, 14, 15. The subject was asked to draw straight continuous lines joining consecutive numbers, i.e., 1 to 2 to 3 etc. A stopwatch was used to monitor the time taken. In the other modified form, the consecutive numbers were so arranged that the child had to make a Z pattern to draw lines joining them.

A Visual Search task was used by Teuber, Battersby, and Bender (1949) to identify visual search deficits following cerebral lesion. The more recent use of this planning marker test (Das, 1980), has involved the use of 12 overhead transparencies which are used in a viewing apparatus which permits accurate timing. Each transparency contains a field of figures, numbers, or letters, and subjects are required to find an instance of a target item, presented in a circle in the centre of the field, and point to its location in the field. The transparencies range in difficulty. The apparatus was so constructed that the subject depresses a button (with his index finger of his preferred hand) in order to view the transparency, and keeps it depressed until he locates the target. Immediately following the sighting of the target embedded in the field, the subject takes his finger off the button to touch the target. The duration for which the button is depressed is search time, and the interval between releasing the button and touching the target is reaction time. In previous research, search time loaded on a 'planning' factor, as it reflects time to scan and make a decision. In the present task administration, subjects were initially given training trials, in which they learned to expect the target items in either a clockwise,

or counterclockwise sequence, in each series of four transparencies.
During test administration, the mean times for the initial two, and
the final two transparencies were calculated for sets of four, again
providing information as to the comprehension and utilization of a
strategy. Both Trail Making and Visual Search were used in the
modified version in our laboratory by Heemsbergen (1980). However,
his subjects were college students, and not children. Both had their
major factor loadings on the planning factor.

Planned Composition was another planning marker task; in this
task, the subject was to write a story based on Card #2 of the
Thematic Apperception Test (Ashman & Das, 1980). There was no time
limit, and stories were scored by averaging the ratings of at least
two teachers at the child's grade level. Criteria used in scoring
included: organization, expression, individuality, wording, and
mechanics (punctuation, capitalization, etc.). The raters assigned
a score of 1 to 7 to indicate their evaluation ranging from good to
poor on each criterion. Their ratings were averaged for each
subject.

Syllogistic Reasoning is not an established test for planning.
However, in the present and other recent studies (Heemsbergen,
1980), we were interested to know if the time required to solve sets
of syllogisms and/or errors made may be sensitive to aspects of
coding and planning. Two sets of eight 3-term syllogisms were given
to each subject, with time for solution of a set recorded by stop
watch. Number of errors were also noted. The first set included
syllogisms such as:

John is taller than Bob

Bob is taller than Dick

Conclusion: John is taller than Dick

True or False

In some examples the adjective "shorter" was used, but in no cases
were conversions required. In set two, the syllogisms were of the
type which required a conversion such as:

Dan is shorter than Jim

Mark is taller than Jim

Conclusion: Dan is shorter than Mark

True or False

One-half of the syllogisms in either set had a 'true' and the other half a 'false' conclusion. Subjects were required to circle one of the choices.

Besides the above tests, all subjects had taken the Edmonton Elementary Reading Test, and the Canadian Cognitive Abilities Test, the latter having verbal, quantitative, and nonverbal IQ components. The reading test was made up of decoding and comprehension, and a subject's score was a percentile within his age group.

A table of means (Table 1) and of F-ratios (Table 2) are provided for inspection. In order to compare the performance of children in Grades 4 and 6, as well as between the reading disabled and normal readers, analyses of variance were carried out for each task. For example, Figure Copying scores of the four groups were subjected to a 2 (Grades 4 and 6) x 2 (normal and disabled readers) analysis of variance. The results for all tests are reported in Table 2 which lists the F-ratios for main effects and interaction. On each test, except in one, the older Grade 6 children did better than the younger Grade 4 children, and the reading disabled were poorer than the normal readers.

The interaction was significant for Visual Search and Syllogism solution time. As the table for mean scores (Table 1) shows, the unusually long Visual Search time of the reading disabled group in Grade 4 was the chief contributing factor to the interaction. It is apparent that the younger reading disabled group was particularly poor in Visual Search. It would be worthwhile to study the development of Visual Search strategies in control as well as reading disabled children to understand the discontinuity noticed here. In syllogistic reasoning time, the interaction is attributable to more or less the same factor – the Grade 4 reading disabled group took much longer, and especially so when compared to the Grade 6 disabled group. An important difference between the groups was obtained in their performance on Serial Recall; the two reading disabled groups were much lower on this task than the controls.

Planned Composition was rated in terms of expression, organization, wordings, mechanics (grammar) and individuality, each on a 7-point scale (rating of 1 was the highest, 7 was the lowest). Only the rating on 'organization' is considered in Tables 1 and 2 because this aspect of composition had been shown in the past to load highly on a planning factor (see Ashman & Das, 1980). It should be noted that the same rater evaluated the protocols of composition of all groups. Three different raters, who were school teachers, were employed, and their ratings were averaged for each subject. The compositions were not identified for the raters in terms of subject or group. As seen in Table 2, both main effects were significant.

Table 1. Mean Scores on Cognitive Tests

	FC	SR	TRM (secs)	VS (secs)	Syll. (secs)	Syll. Error	Comp.
GRADE 4	19.0	50.8	6.88	3.42	115.27	2.49	5.17
GRADE 6	24.4	59.9	4.88	2.55	112.90	1.74	4.40
LD:CA GRADE 4	15.6	30.6	7.45	5.54	179.83	3.17	6.22
LD:CA GRADE 6	19.1	38.0	6.70	3.38	118.37	3.00	6.00

Legend: LD:CA = Learning Disabled and Chronological Age
         FC = Figure Copying
         SR = Serial Recall
         TRM = Trail Making
         VS = Visual Search
         Syll. = Syllogisms
         Comp. = Composition

Table 2. Summary of 2 x 2 analyses of variance: $\underline{F}$-ratios.
df = 1,114. Only significant values are given.

	Grades 4/6	Normal/RD	Interaction
Figure Copying	14.72	15.16	NS
Serial Recall	109.65	25.02	NS
Trail Making	11.28	8.44	NS
Visual Search	22.40	23.71	4.31
Syllogism Time	7.76	9.34	6.65
Syllogism Error	NS	13.30	NS
Composition (organization)	27.19	4.01	NS

One may raise the question of the occurrence of Type 1 error
because of the seven 2 x 2 analyses of variance which were performed. Perhaps a multivariate analysis of variance would be pre-
ferable. Further, we have done analyses of covariance in order to
partial out the effect of IQ on test performance. Thus, a multivariate analysis of covariance followed by further tests to tease
out group differences on each test should be carried out. However,
the F-ratios for main effects were significant below .01 level,
which offers a hedge against Type 1 error. Besides, all the group
differences were expected. The only one which did not become
significant was syllogism error scores for Grades 4 and 6. In
analyses of covariance, all results agreed with those in analyses
of variance. Thus, it would appear that the F values were not over-
estimated to an extent which would result in obtaining statistical
significance where there was none.

From the point of view of tasks which had a 'memory' component,
it is clear that Serial Recall, and to a minor extent, Trail Making
can be thus identified. In all other tasks which included Figure
Copying, Visual Search and Syllogisms, it would be reasonable to
assume that there was little load on memory. From another classi-
ficatory point of view, verbal versus nonverbal, Figure Copying and
Visual Search were nonverbal relative to Serial Recall, Syllogisms
and Composition. The reading disabled groups, however, were less

competent on tasks with and without a memory component as they were
on the verbal and nonverbal tasks.

In summary, the analysis of variance showed that the reading
disabled groups were poorer in planning tasks, and on the two coding
tasks. However, contrary to our expectation, they were not poorer
in planning than they were in coding. Further, the disabled group
took longer to do syllogisms.

Since Das, Leong, and Williams (1978) had established the
inferiority in performance of the disabled group who were matched on
nonverbal IQ with the normal readers, we do not expect that the
present results are due to IQ differences which exist between the two
groups. However, one may argue that while Leong's results were
unrelated to IQ, ours might not be so. It is difficult to see how
this possibility can be entertained; however, we have dealt with the
IQ difference by using analyses of covariance.

The reading disabled group in the present study had mean nonverbal IQs of 84.93 (Grade 4-RD) and 90.80 (Grade 6-RD) compared to
101.16 and 105.61 for normal readers as indicated by the performance
scale of the 'basic skills' test, and the difference was significant. Therefore analyses of covariance were done, partialling out
the effect of IQ. Correlations between IQ and performance on the
cognitive tasks were available from the computer program for
analysis of covariance. Some of the high rs were obtained with
Serial Recall (0.50), Figure Copying (0.46) and error scores of
Syllogistic Reasoning (0.57), and Composition (0.35). But the
results of the analyses of covariance were essentially congruent
with those of analysis of variance without any exception.

How far backward was the test performance of the reading disabled when compared to the normal readers? Simply the means are
shown in Table 1 in order to estimate the extent of backwardness.
It is remarkable to find consistently that in 5 of the 7 tests, the
Grade 6 reading disabled and the Grade 4 normal readers had almost
identical scores, and in the remaining tests, the Grade 6 reading
disabled were much lower. These results tend to suggest the existence of a general cognitive processing deficit along with a
specific deficit for verbal processes in the reading disabled
samples. The extent of general deficit seems to be equivalent to
the extent of backwardness in reading.

How sensitive are the coding and planning tasks to differences
in reading competence within the normal range? We have attempted
to answer this question by dividing the Grade 4 normal readers into
those who scored above the median and those who scored below on the
decoding test. Similarly, they were divided on the basis of their
comprehension scores. Grade 6 normal readers were similarly divided

into above and below median groups.   There were 22 children in each
group.   The results of Welch's t tests are summarized in Table 3.
Only in very few cases, variances were not homogeneous; therefore
Welch's t values were in agreement with the 2-tailed t tests for
almost all mean comparisons.   We have expected from our previous
work on reading that successive processes (as exemplified here by
Serial Recall) were salient for decoding (Cummins & Das, 1977) for
samples of children in early grades, and that both successive and
simultaneous processes contributed to comprehension.   Grade 4 children are not beginning readers, but nevertheless, in decoding,
successive processes are expected to be used.   It was not surprising, therefore to find that t for Serial Recall was highly significant when decoding was the criterion variable, and t values for both
Serial Recall and Figure Copying were significant for comprehension.
For the Grade 6 sample, the above tasks were salient for decoding,
but only Figure Copying was of importance for comprehension.   Going
along with the coding and planning model to look at the tasks, one
notices that the 'planning' tasks are hardly relevant for decoding
or comprehension, except for the Composition task.   The latter requires verbal skills as does Syllogistic Reasoning.   Indeed all the
tests in Table 3 could be looked upon as verbal except Figure
Copying.   A different sort of experiment has to be designed if one
wishes to study the relative contribution of verbal and nonverbal
contents in coding and planning tasks to skills in decoding and comprehension.

     In discussion, we wish to mention the following points.   First
is the general point in the choice of tasks.   Although one could
have selected tasks on some atheoretical ground, the present choice
was justified by Luria's organization of cognitive functions (Luria,
1966) as operationalized by us (Das, 1973a, 1973b; Das, Kirby, &
Jarman, 1979).   Apart from the neuropsychological underpinnings of
the theory which could well be ignored by some readers, the model
gave one the initial advantage of covering basic cognitive processes, coding and planning, which operate on perceptual, memorial and
conceptual information.   The experiment showed that whereas the
disabled readers at Grade 6 are about two years behind in the simultaneous and planning tasks, they are excessively backward in the
only successive task which was used.   This task required verbal
successive processing.   A purely nonverbal instance of successive
processing is hard to find, but one with reduced verbal mediation,
such as learning the sequence of "O + + -" randomly repeated for
three cycles could be used.   Would the disabled readers show
relatively less incompetence on such a task than they did in seri-
ally recalling word lists?

     The disabled readers in Grade 4 were markedly behind the con-
trol children in the same grade, and seem to be as backward as the
Grade 6 disabled readers were in relation to their controls.

Table 3. Comparison of good (N=22) and poor (N=22) readers (above
          versus below median) on cognitive tasks. Welch t-test
          adjusted for unequal variance. Only significant
          differences are mentioned. *P < .05; **P < .01.

	DECODING	COMPREHENSION
GRADE 4	Serial Recall**	Figure Copying*
	Syllogisms (Errors)*	
	Planned Composition*	Planned Composition**
GRADE 6	Figure Copying**	Figure Copying*
	Serial Recall**	
	Syllogisms (Errors)**	Syllogisms (Errors)**
	Planned Composition**	Planned Composition*

Decrement in performance from Grade 6 to 4 among control groups is
parallel to the decrement in the disabled groups. Therefore, insig-
nificant interactions are found for all tests except in the case of
Visual Search; in that task, the slope is steeper for the disabled
groups (compare the means in Table 1).

   Arising out of the present study, there is a need to examine
visual scanning in young reading disabled children. One should
check, in a subsequent study if reading disabled children at Grade
4 and younger, experience specific difficulties in strategies for
visual scanning. The reading disabled child is often found to be
slow in vocalization in addition to Serial Recall. All of these
behaviors are instances of successive processing. It would be
worthwhile to design tasks which are predominantly nonverbal and
successive, and then contrast the performance of the reading dis-
able group and their controls.

   In conclusion, the disabled readers have appeared to be generally backward in coding and planning functions and specifically
defective in some of the tasks which measure these. There was some
suggestion that Visual Search strategies interact with development.
The skilled and less skilled readers are distinguished only by some
of these tasks, and most clearly in planful composition.

Prediction of reading

Next, we would like to summarize a study by Mulcahy in which he found that Digit Span and Figure Copying were the best predictors of competence in reading and spelling in elementary school children. He used a regression analysis to demonstrate that these two basic cognitive tests of successive and simultaneous processing were better predictors of reading and spelling than other tests which were closer to reading, such as, story-recall. The subjects were between the ages 8 and 12 years. They were drawn from regular and learning assistance classrooms in a suburban school district near Edmonton.

Teachers nominated a pool of subjects for each of the three levels of reading proficiency. Subjects were selected from the subjects nominated on the basis of file data including chronological age, IQ, and reading scores from the Metropolitan Achievement Test and Canadian Test of Basic Skills. Prior to experimental testing, the Wide Range Achievement Test (WRAT) reading and spelling tests were administered to all subjects. Some subjects were excluded because the WRAT reading level criterion and teacher's nomination did not place the child at the same level of reading proficiency.

The overall study sought to delineate different aspects of coding, planning, memory, metamemory, and their relationships to age and general level of reading proficiency. For purposes here, only data directly related to coding and reading will be discussed.

234411

In the first set of analyses, children were divided into four groups on the basis of their scores on Figure Copying (marker test for simultaneous) and Digit Span (marker test for successive) such that those who scored above the median on both, below the median on both, and above one, but below on the other could be identified. The groups thus generated by a double-median split were compared on vocabulary, spelling, and story recall (number of propositions recalled). Analysis of variance (high and low on simultaneous versus successive tasks) showed significant effects for simultaneous processing when vocabulary and spelling scores of the children were considered - those above the median in Figure Copying had higher scores on vocabulary and on comprehension, whereas no relationship was found for groups divided on Digit Span. However, the latter, the high and low successive groups, were differentiated in terms of the time taken to read the story to be recalled.

In the next set of analyses, multiple regression was used. The given predictors were Figure Copying, Digit Span, Semantic Recognition and two measures of Story Recall (propositions recalled and time to read). Spelling competence was best predicted by Digit Span

followed by Figure Copying. Reading competence was similarly predicted. It is worth noticing that these predictor variables are not
as directly related to reading as is story recall.

The significant question here is whether the reading disabled
children lack the capacity for successive processing or they have
the capacity, but do not use it for some reason in reading. As far
as we can conclude from our research (Das & Cummins, 1982), the
problem seems to be in the utilization of successive processes. It
seems that a weakness in their control processes is displayed in
their non-utilization of successive processes, as well as their
knowledge base in vocabulary and comprehension. The defect is thus
a composite one consisting of successive processing and planning.
The reading disabled children lag behind in effective organization
of their successive coding capacity. There seem to be strong indications that whatever sequential processing deficits that the reading disabled might have, it is compounded by inconsistent and inadequate utilization of sequential processing.

COMPETENCE, PROCESSES AND REMEDIATION

One must clearly understand the nature of competence in reading
and comprehension. The key questions here are both methodological
as well as substantive. In comparing the reading disabled children
with normal children, the scientific literature has documented in-
numerable reports of meticulously carried out experiments in which
the reading disabled was found to be inferior to the average reader
in one or more cognitive tasks. The finding would not come as a
surprise to teachers, who may wonder if that was not obvious. It is
obvious that the reading disabled child is inferior in tasks that
involve reading or activities which are closely associated with
reading and language, but what is not so obvious is their inferiority in tasks such as Figure Copying or Visual Search, which do not
involve reading skills. At least the connection between the two
classes of skills is not obvious. Some researchers, such as Bradley
and Bryant (1978), have matched reading disabled children with
average readers on reading age, that is on scores on word recognition or on some other reading tests. In spite of such matching, the
reading disabled child is still found to do more poorly than the
younger non-disabled child on certain tasks. In Bradley and
Bryant's (1978) case, the task involved organization of words which
are presented auditorially to the children.

An honest attempt to understand competence in reading leads one
to considering underlying processes. The majority of authors in
this book have examined processes associated with reading disability or reading ability and have made significant contributions.

However, a general approach to processes which we would like to
suggest is to start with broad cognitive processes, such as coding
and planning. Subsequently relate them in a heuristic manner
through mainly correlational analysis and factor analysis to reading
competence; but then to abandon these procedures and get to an
experimental analysis of the tests and tasks themselves in order to
identify the components of cognitive processes. One such example
might be a phonological coding task. What does phonological coding
involve? In an attempt to answer this, one may design experiments
on phonological coding and relate the results of such experiments
to individual differences in the basic cognitive processes as
identified, for example, in our model. In this approach there is a
place for both the top-down and bottom-up processes and room for
both chronological and reading age-matched designs. The approach
is essentially a dialectical one; starting with broad cognitive
processes, proceed to an analysis of specific components of reading.
Knowledge gained from such analysis flows back to the conceptualization of the broad processes.

As we do all this, that is, measure competence and try to
understand it in terms of processes, the task will not be complete
unless our results provide not only a base, but a direction for
remediation. Perhaps we already have some directions for constructing rational remediation procedures. First of all, we are quite
aware of the fact that specific problems in coding may lead to general problems in reading. As the work of O'Connor and Hermelin
(1978) on handicapped populations has shown, a broken link may
weaken the whole chain. Thus a need for a complete cognitive mapping of a reading disabled child in a clinical situation is indicated. For some children, repairing the broken link may restore the
strength of the child in reading.

It seems to us that there are essentially two promising approaches to remediation, global and specific. Cognitive mapping as
recommended above, can reveal general process deficits, such as in
successive processing. It may further demonstrate that a reading
disabled child has the capacity for such processing, but does not
know how to utilize it in an optimal fashion. If this be the case,
a remediation program might best be devised to facilitate successive
processing and its utilization. We have attempted to develop such a
program and have applied it with some success (Brailsford, 1981;
Kaufman, 1978; Krywaniuk, 1974). The underlying logic of using
global remediation is that by strengthening the processes, skills
required for reading will also be strengthened.

A summary of the three attempts at remediating reading backwardness by global process training is given here. The reader can
thus have an idea of the approach and compare it with specific
training which will be described in a subsequent section. The first

study in which general training for successive processing was
attempted was carried out on Canadian Native children who lived on
a reserve (Krywaniuk, 1974). All children went to a school on the
reserve and were in Grade 3. All were found to be poor in language
achievement tests, but about one-third of the children in the Grade
3 class were identified as particularly deficient in reading and
language. These children were included in the intervention study.
The group had 20 children in maximum intervention sessions lasting
for nearly 15 hours and 20 children in a minimum intervention session of 3 hours duration. Intervention consisted of training children on a set of tasks which required successive processing. All
tasks were easily available, and did not have any specific relation
to the content of reading tasks. All required some amount of verbalization, but almost no reading.

Sequence Story Boards will serve as an example of the intervention exercises. It consists of three separate stories, each
having 12 removable pictures that could be arranged to tell a story,
which are about a trip to the zoo, grocery store, and building a
house. The pictures were to be arranged by the child in such a way
that it would tell a good story. As the child told the story after
arranging the pictures in a row of four, the inconsistencies, if
any, in his arrangement were pointed out, and he was allowed to
correct them. A number of tasks were given individually, but there
were some tasks which were administered in groups of 3 or 4 children. Film strips which required visual and spatial discrimination
of dots located in a quadrant would be one of the tasks -- the child
was asked to point out "the dot to the right and above," "to the
left and above" (there was no such dot) etc. The results of training were shown not only in the improvement of performance on successive tasks, but also in word-recognition. Of course, significant
gains were achieved by only the group which received the 15 hours
intervention program (Krywaniuk, 1974).

The Native children were not dyslexics, but showed gross underachievement in reading. Their verbal intelligence was lower than
average, but in performance intelligence, they were comparable to
Grade 3 children in an Anglo school. Underachieving children in a
regular Anglo school were the subjects for the next intervention
study by Kaufman (1978). Like Krywaniuk (1974), he selected tasks
which required mainly successive processing. Verbalization of what
the child was doing was encouraged, and was perhaps emphasized much
more than it was in the previous study. Nine separate tasks and
seven film strips were utilized for training; only some of these
materials were identical to Krywaniuk's (1974). Again, the treated
group improved in word-attack skills over the nontreated group.
The essence of training comprised labeling and verbalization along
with teaching successive processing.

The last study in this series selected 9- to 12-year-old children with learning disabilities; these children had normal IQs, but were significantly backward in comprehension (Brailsford, 1981). Twelve children were taken out of their regular remediation class for a total of 15 hours and given exercises on successive and on simultaneous tasks. An equal number of disabled children in the remedial class acted as controls. As expected, the treated group significantly improved in their reading level. The children in this group also showed a trend towards increased production of inferences, summaries, and syntheses (Brailsford, 1981). Some of Brailsford's tasks were quite innovative, especially the simultaneous tasks such as Tracking and Categorization.*

The other approach, specific training, complements the first one. The strategy here is to identify those reading-related tasks in which the performance of reading disabled children is poorer than those of average readers who have an equivalent reading age. Bradley and Bryant (1978) have found such a task: Children were asked to pick the odd word in a set of four words, three of which had either the same initial, middle or end phoneme. Seymour and Porpodas (1980) also mention several classes of tasks in which the performance of the backward reader was inadequate. The reading disabled children were not sensitive to orthographic regularity, that is, when asked to judge if orthographically regular nonwords such as SLART and SRALT were the same or different, the disabled readers took longer to decide, or made more errors. The authors suggest that backward readers have special difficulties in verbal tasks which require spatial organization and temporal ordering. All of the abilities identified by Seymour and Porpodas (1980) can be accommodated within the simultaneous-successive framework.

There exist several other linguistic principles which should be considered in constructing reading related remediation programs. Training could involve some awareness of language rather than specific ways in which language is written. For instance, consider phonological coding. The skilled readers not only pronounce words that they know, but also words that they do not know, because they can use rules of phonological coding. Exposure to pronounceable nonwords should facilitate the unobtrusive learning of rules

---

* Tasks involving tracking and categorization of shapes were adapted from L. A. Venger and V. V. Kholmovsko. The diagnostics of intellectual development in pre-school children. Moscow: Pedagogica, 1978 (published in Russian). A task involving matrix numbers was adapted from the training battery of D. Kaufman. Strategy training and remedial techniques. In R. F. Jarman and J. P. Das (Eds.), Issues in developmental disabilities. Ann Arbor, Michigan: University Microfilms International, 1980.

associated with grapheme-phoneme correspondence. In order to test
the disabled reader's skill in phonological coding, direct methods
as well as indirect ones could be used. The direct ones are simply
tasks which require the reader to pronounce words and pronounceable
nonwords in a pre-post design in order to assess the effect of
training. Indirect procedures could be ingenious. For instance, a
task used by Coltheart (1978) involved the recognition of words and
nonwords -- the subject was to judge if the item was a word. Some
of the nonwords were pseudohomophones (e.g., BLUD), others were
merely pronounceable; none had any meaning, which made them nonwords. Skilled readers, who normally use phonological coding in
these circumstances, will be relatively slower than disabled readers
in recognizing a pseudohomophone to be a nonword than recognizing an
ordinary nonword as such. That is, although the skilled readers
will be usually faster in word recognition, their advantage would
diminish for pseudohomophones. If this paradoxical effect is
verified to be true, then the procedure can be used to judge the
improvement in disabled readers at the end of a training period for
phonological coding -- the should show the paradoxical delay as they
acquire the habit of phonological coding!

Specific remediation of reading disability has been attempted
in an increasing number of studies. Unlike earlier studies, some
of the very recent ones have been well-controlled. We wish to
mention briefly three of them, published in 1981. In Germany,
Scheerer-Neumann (1981) has developed a training program in which
backward readers are taught to segment words into syllables. It was
observed that these children do not perceive the whole word. In
fact, they do not go beyond the first few letters. Therefore, as a
compensatory exercise, children are first given several practice
sessions to find the vowels in written words, then to mark the
boundaries of syllables. Next, they are asked to read the words
syllable by syllable. In the final phase of training, they are
encouraged to read the word as a whole. Apparently, backward
readers can recognize a syllable and can guess what the word is when
it is presented in fragments. They know the rules of syllable formation, but they cannot adequately analyze the stimuli to verify the
rule. Results of training were seen in a significant reduction of
errors during oral reading.

Decoding strategies were also taught by Fayne and Bryant (1981)
to a large sample of backward readers. They examined the relative
benefits of teaching or coding by adopting one of three techniques:
teach initial bigrams and the final consonant (co- p), or final
bigram but initial consonant (c-op), or letter by letter analysis
(c-o-p). Initial bigram teaching was most effective as judged by
the children's ability to read new words. These findings suggest
that the usual method of teaching word-reading through rhyming or
letter by letter analysis is wrong - at least that is the strong

conclusion arrived at by the authors. Apart from the question of
what to teach the reading disabled child, the results in regard to
the efficacy of emphasizing initial bigrams agree with Bryant and
Bradley's observation (see their chapter in this book) - they asked
children to detect the organization of words which had similar
initial alphabets, middle or last alphabets. The reading disabled
children experienced their greatest difficulty in grouping the words
with similar initial alphabets.

Syntax is difficult to grasp for most of the backward readers.
Like the decomposition of words into successively occurring syllables, or even bigrams, the child must use successive processing (and
some simultaneous processing) in order to understand syntax. A
training program in syntax was designed by White, Pacarella, and
Pflaum (1981). They gave packets of words that could be combined
into sensible English sentences to backward readers, and succeeded
in improving the children's comprehension by using the cloze technique.

Besides these recent examples of remediation, one should take
into account Bryant and Bradley's longitudinal study (see their
chapter in this book). Bradley has developed remedial programs
which have been effective in improving reading. These programs can
be administered by school teachers. A good example is her kit
"Sound Pictures" marketed by Macmillan Education Ltd. in London.
The object of this teaching aid is to make children aware of grouping words together which sound alike (e.g., hen, pen, men). She has
picture cards corresponding to these words which are given to the
child for grouping. In summary, the above remediation programs
which are specifically related to reading skills seem to work as
well as the global ones which improve general cognitive processes.
By combining these two kinds of programs, one should be able to
produce maximum improvement.

The relationship between global and specific remediation is to
be understood in terms of the relation among broad cognitive processes, reading-specific processes, and reading skills. Language
awareness and phonological coding were the objects of improvement
in the specific programs. These have a closer relationship to
reading performance than simultaneous-successive processing. What
is assumed here is that the path of influence of cognitive processes on reading performance may run through linguistic processes.
The reason for the improvement in reading performance following
training programs on successive processing (Kaufman, 1978;
Krywaniuk, 1974) may be the improvement of knowledge of linguistic
principles including metalinguistic awareness. There is no evidence
at present which charts the path of influence. But a study by Leong
(1981) has begun to indicate, through path analysis, the relation
between simultaneous-successive processes, language awareness and
word recognition; his results seem to support our assumption.

Perhaps one should speculate on how enduring are the changes brought about by remediation. A related question is how easily could the skills acquired during training be transferred to other situations. Remediation entails imparting knowledge and building up habits of reaction. Knowledge acquired in one situation should be applicable in similar situations; this is easily achieved. But something more than mere transfer to a cognate situation is required to make knowledge useful - it is flexibility. In terms of information processing theory, knowledge should be accessible outside the context in which it was acquired. Cross-indexing, which is the ability to index information about the recent past so that it could be recalled and used in more than one way (Rabbitt, 1981) seems to be the critical prerequisite for transfer of knowledge. A successful remediation program should ensure flexibility as much as possible.

The question of maintenance of the effect of remediation has been overly discussed. It is simply concerned with how newly learned habits can be retained often in the face of previously established competing habits. Interference from established habits has to be resisted, and spontaneous regression has to be prevented. Two ways for achieving these are overlearning and providing incentives. Fast remediation is a contradiction in terms if one desires that its effects endure for a long time. The new skills must be learnt to an extent of becoming automatic, a part of one's second nature. In the reading behavior of reading disabled children, lack of automaticity is commonly noticed, and slowness in comprehension is attributed to absence of automaticity. Thus overlearning is the key to developing automaticity. But overlearning becomes a chore if carried out in artificial and uninteresting situations in spite of the standard incentives and rewards of the laboratory. Imagine how we have learnt vital skills for not only surviving, but prospering in our environment. The key is to maintain interest in learning which naturally occurs in the life of an individual. High motivation accompanies activities learnt through social interactions. For a child, interesting interactions with adults are naturally desired, and most of a child's learning probably occurs during verbal intercourse with adults (Vygotsky, 1963). Here is then a lesson for educators who are engaged in remediation.

In concluding our observations on remediation, it seems reasonable to suggest that remediation procedures following from mapping the cognitive processes of the reading disabled child will work. Such programs will be guided by a theoretical rationale such as the one we have proposed, and must be designed to simulate the social conditions in which habits are learned.

References

Armitage, S. G.   An analysis of certain psychological tests used for the evaluation of brain damage. Psychological Monographs, 1946, 60, No. 1 (Whole No. 277).

Ashman, A. F., & Das, J. P.   Relation between planning and simultaneous-successive processing. Perceptual and Motor Skills, 1980, 51, 371-382.

Barron, R. W.   Visual and phonological strategies in reading and spelling. In U. Frith (Ed.), Cognitive processes in spelling. N.Y.: Academic Press, 1980.

Bradley, L., & Bryant, P. E.   Difficulties in auditory organization as a possible cause of reading backwardness. Nature, 1978, 271, 746-747.

Brailsford, A.   The relationship between cognitive strategy training and performance on tasks of reading comprehension within a learning disabled group of children. Unpublished Master's Thesis, Department of Educational Psychology, University of Alberta, Edmonton, Canada, 1981.

Coltheart, M.   Lexical access in simple reading tasks. In G. Underwood (Ed.), Strategies of information processing. London: Academic Press, 1978.

Cummins, J., & Das, J. P.   Cognitive processing and reading difficulties: A framework for research. The Alberta Journal of Educational Research, 1977, 23, 245-255.

Das, J. P.   Planning: Theoretical considerations and empirical evidence. Psychological Research, 1980, 41, 141-151.

Das, J. P.   Cultural deprivation and cognitive competence. In N. R. Ellis (Ed.), International review of research in mental retardation (Vol. 6). New York: Academic Press, 1973a.

Das, J. P.   Structure of cognitive abilities: Evidence for simultaneous and successive processing. Journal of Educational Psychology, 1973b, 65, 103-108.

Das, J. P., & Cummins, J.   Language processing and reading ability. In K. D. Gadow and I. Bailer (Eds.), Advances in learning disabilities (Vol. 1). Greenwich, Conn.: JAI Press, 1982.

Das, J. P., Kirby, J. R., & Jarman, R.   Simultaneous and successive cognitive processes. New York: Academic Press, 1979

Das, J. P., Leong, C. K., & Williams, N.   Relation between learning disability and simultaneous-successive processing. Journal of Learning Disabilities, 1978, 11, 17-23.

Fayne, H. R., & Bryant, N. D.   Relative effects of various word synthesis strategies on the phonics achievement of learning disabled youngsters. Journal of Educational Psychology, 1981, 73, 616-623.

Frith, U.   Experimental approaches to developmental dyslexia: An introduction. Psychological Research, 1981, 43, 97-109.

Heemsbergen, D. S.   Planning as a cognitive process. Unpublished Doctoral Dissertation, Department of Educational Psychology, University of Alberta, Edmonton, Canada, 1980.

Ilg, F. L., & Ames, L. B.   School readiness: Behavior tests used
    at the Gesell Institute.   N.Y.: Harper & Row, 1964.
Kaufman, D.   The relationship of academic performance to strategy
    training and remedial techniques: An information processing
    approach. Unpublished Doctoral Dissertation, Department of
    Educational Psychology, University of Alberta, Edmonton, Canada,
    1978.
Kirby, J. R., & Das, J. P.   Reading achievement, IQ and simulta-
    neous-successive processing. Journal of Educational Psychology,
    1977, 69, 564-570.
Krywaniuk, L. W.   Patterns of cognitive abilities of high and low
    achieving school children. Unpublished Doctoral Dissertation,
    Department of Educational Psychology, University of Alberta,
    Edmonton, Canada, 1974.
Leong, C. K.   Reading and its breakdown - Role of language aware-
    ness. First IARLD Conference, Utrecht, 1981.
Leong, C. K.   Spatial-temporal information processing in disabled
    readers. Unpublished Doctoral Dissertation, Department of
    Educational Psychology, University of Alberta, Edmonton, Canada,
    1974.
Lesgold, A. M., & Perfetti, C. A.   Interactive processes in
    reading. Hillsdale, N.J.: Lawrence Erlbaum Associates, 1981.
Luria, A. R.   Higher cortical functions in man. New York: Basic
    Books, 1966.
Nelson, H. E., & Warrington, E. K.   An investigation of memory
    functions in dyslexic children. British Journal of Psychology,
    1980, 71, 487-503.
O'Connor, N., & Hermelin, B.   Seeing and hearing and space and
    time. London: Academic Press, 1978.
Pavlidis, G. Th.   Sequencing, eye movements and the early objective
    diagnosis of dyslexia. In G. Th. Pavlidis and T. R. Miles
    (Eds.), Dyslexia research and its application to education.
    Chichester, England: John Wiley, 1981 (chapter 8).
Perfetti, C. A., & Goldman, S. R.   Discourse memory and reading
    comprehension skill. Journal of Verbal Learning and Verbal
    Behavior, 1976, 14, 33-42.
Rabbitt, P.   Human ageing and disturbances of memory control
    processes underlying "intelligent" performance of some cognitive
    tasks. In M. P. Friedman, J. P. Das, and N. O'Connor (Eds.),
    Intelligence and learning. N.Y.: Plenum Press, 1981.
Reitan, R. M.   The relation of the Trail Making Test to organic
    brain damage. Journal of Consulting Psychology, 1955, 19, 393-
    394.
Scheerer-Neumann, G.   Prozessanalyse der Leseschwache. In R.
    Valtin, U. O. H. Jung, and G. Sheerer-Neumann (Eds.),
    Legasthenie in Wissenschaft und Unterricht. Wissensch oftliche
    Buchgesellschaft, Darmstadt, 1981.
Seymour, P. H. K., & Porpodas, C. D.   Lexical and non-lexical
    spelling in dyslexia. In U. Frith (Ed.), Cognitive processes
    in spelling. London: Academic Press, 1980.

Shankweiler, D., Liberman, T. Y., Mark, S. L., Fowler, C. A., & Fischer, F. W. The speech code and learning to read. _Journal of Experimental Psychology, learning, and Memory_, 1979, _5_, 531-545.

Spreen, O., & Gaddes, N. H. Developmental norms for 15 neuro-psychological test ages, 5 to 15. _Cortex_, 1969, _5_, 171-191.

Teuber, H. L., Battersby, W. S., & Bender, M. B. Changes in visual searching performance following cerebral lesions. _American Journal of Psychology_, 1949, _159_, 592.

Torgesen, J. K. Conceptual and educational implications of the use of efficient task strategies by learning disabled children. _Journal of Learning Disabilities_, 1980, _13_, 19-26.

Torgesen, J. K., & Houck, G. Processing deficiencies in learning disabled children who perform poorly on the Digit Span task. _Journal of Educational Psychology_, 1980, _72_, 141-160.

Underwood, G. (Ed.), _Strategies of information processing_. London: Academic Press, 1978.

Vellutino, F. R. Alternative conceptualizations of dyslexia: Evidence in support of a verbal deficit hypothesis. _Harvard Educational Review_, 1977, _47_, 334-354.

Vygotsky, L. S. _Thought and language_. Cambridge, Mass.: MIT Press and Wiley, 1963.

White, C. V., Pascarella, E. T., & Pflaum, S. W. Effects of training in sentence construction on the comprehension of learning disabled children. _Journal of Educational Psychology_, 1981, _73_, 697-704.

THE USE OF RATIONALLY DEFINED SUBGROUPS IN RESEARCH ON LEARNING DISABILITIES[1]

Joseph K. Torgesen

Florida State University
Tallahassee, Florida, U.S.A.

Many of the chapters in this volume are critical of both current theory and research on learning disabilities. Although this critique does not deny that the last 20 years of research have produced many new and useful ideas about LD children, it does suggest that we must begin to change our methods in fundamental ways if we are to continue to make progress in understanding some of the most difficult issues in the field. This paper presents a model for doing research on learning disabilities that is not radically different from current procedures, but which does provide a partial solution to some of the problems in most current research efforts. To provide a context for this model, we will first consider some of the more pressing needs of the field that good research may help to solve.

CRITICAL NEEDS IN THE STUDY OF LEARNING DISABILITIES

The needs to be discussed here are only relevant if one's goal is the continued maintenance of a distinction between learning disabled children and children who fail in school for a variety of other reasons. Given that ultimate goal, one of the most critical needs of the field of learning disabilities is the establishment of a clear sense of identity. One of the core ideas of the field is that some children fail to learn in the classroom because they process information differently (deficiently) when compared to children who learn at a normal rate. This idea has a

---

[1] Preparation of this manuscript was partially supported by a grant from the Thrasher Research Fund, Salt Lake City, Utah.

111

strong conceptual base (Torgesen, 1979), but it has proven very
difficult to obtain a consensus about the specific processing
skills which contribute to the problems of learning disabled
children. Given some current critiques that suggest LD children
are not much different from other underachieving students
(Hallahan & Kauffman, 1976; Ysseldyke, Algozzine, Shinn, & McGue,
1979), there is a clear need to establish the validity of the
central defining feature of the field by demonstrating reliable
and clearly defined processing difficulties in LD children.

Another critical need in the field of learning disabilities
is to establish the manner in which processing disabilities actually affect performance on academic tasks. The assumption is
normally made that academic tasks like learning to read require
some processing activity which certain LD children find very difficult. When psychometric tasks show a LD child to perform deficiently in certain areas like perceptual-motor integration, visual
perception, or short-term memory, the diagnostician usually assumes
that these "processing deficits" underly the child's poor reading
performance. However, the coincidental occurrence of a reading
problem and a processing deficit is not evidence that one causes
the other, or that they are related in any important way.

The task of developing reliable data and theory to establish
a link between cognitive processing deficits and specific forms
of academic failure is an important one for future research. Recent information processing research on tasks like reading (Reber
& Scarborough, 1977; Resnick & Weaver, 1979) has begun to identify
the component processing skills required by complex academic tasks.
Research on the processing skills of learning disabled children
must be sufficiently analytic to allow identification of component
processes that may be shared between diagnostic tasks and reading
or other academic tasks.

Finally, we need to develop remedial strategies that are
responsive to individual differences in processing skills. At
present, there is little evidence that techniques which attempt
to take the processing deficiencies of LD children into account
are any more effective (or as effective, for that matter) as
approaches which assume that all children learn similarly (Smead,
1977). As Wozniak (1979) has pointed out, the failure of process
oriented remedial techniques to prove their effectiveness is not
necessarily evidence against the basic concepts involved, but may
be more related to technical inadequacy of current approaches in
several areas. For example, we really do not have an adequate
description of which processing skills are important for effective
learning in the classroom. The current technology of measurement
of processing deficiencies is also inadequate (Torgesen, 1979).
Sound analytic research on the processing problems of LD children

should not only produce better measurement techniques, but should also lead to better understanding of which processing skills are crucial for learning. Although knowing which processes are defi- cient is clearly not the same as knowing how to remediate them (Resnick, 1979), understanding why children fail tasks certainly provides useful information from which to begin the development of remedial programs.

Most experimental work to date has addressed the need to identify the processing deficiencies which characterize LD chil- dren. This work has generally consisted of isolated studies that have compared the performance of LD and normal children on tasks thought to require certain processing skills. Most of these stud- ies are not followed up systematically by subsequent experiments (Torgesen & Dice, 1980) and thus they usually raise more questions than they answer. There has also been a significant amount of research on various remedial approaches, but almost no research focused on the link between processing deficiencies and academic failure (other than strictly correlational studies) has been done. We know of no research programs that have addressed all three goals in a sequential, coherent manner. Thus, we have very little useful information about the relationship between processing de- ficiencies, academic deficits, and remedial procedures.

Of course, there have been a substantial number of studies which assessed processing deficiencies by a standard psychometric device and then provided remediation of the identified deficiencies. Such studies have failed to demonstrate the utility of assessing processing deficiencies as a preliminary step in remediation. However, they are all seriously weakened by lack of knowledge of the processes that most standard tests actually measure. Thus, in order to accumulate more accurate information about the way various processing deficiencies interact with certain remedial procedures, these procedures have to be implemented as part of a research program which has previously established the precise na- ture of the processing deficits involved. One important step in such a research program would be to show that the processes being studied play an important role in the academic failure of LD children.

THE PROBLEM OF SAMPLE HETEROGENEITY

An assumption shared by most modern researchers in learning disabilities is that children categorizable under this lable repre- sent a heterogeneous population. In other words, a given sample of LD subjects is likely to contain children who have failed in school because of a variety of different information processing problems. The heterogeneity of most samples of LD children compounds the difficulty of demonstrating effectiveness for remedial procedures that focus on specific processing deficiencies. Since most randomly selected groups of LD children are likely to contain only a few children with the processing deficit in question, a given remedial procedure may be simply irrelevant for most of the sample. In order to fairly test the effectiveness of process oriented remedial approaches, we need to have reliable means of specifying before treatment which children they are most likely to benefit.

Unfortunately, although most researchers would agree that randomly selected samples of LD children are heterogeneous with regard to the reasons for their failure in school, few studies actually take account of this heterogeneity in their design. For example, Torgesen and Dice (1980) examined almost 90 studies reported in major education-psychology journals over the past three years and found that none of them employed any system to reduce the heterogeneity of their samples of LD children. Thus, virtually all of the current research is being conducted on heterogeneous samples of LD children. This state of affairs not only makes it difficult to verify the utility of process-oriented remedial approaches, but also clouds the basic data on processing deficiencies. Although many studies show potentially interesting patterns of processing deficiencies in LD children, these results are not clinically useful because we have no way of specifying to which LD children the results of a given study apply.

The ideal solution to the problem of heterogeneity in LD samples would be a systematic taxonomy of LD subtypes. Such a taxonomy would describe a finite number of LD subgroups that were homogeneous with regard to the major variables contributing to failure in school. There are beginning efforts to derive such a taxonomy using statistical techniques such as cluster analysis on large samples of LD children (e.g. Satz & Morris, 1980). Not only is this technique very difficult to apply, but also the particular clusters identified depend critically on the types of data collected and the level of coherence that is arbitrarily demanded of each cluster. Thus, correct application of this technique will require an extensive and coordinated effort to cross-validate clusters and develop a battery of appropriate assessment techniques to generate data for analysis. With these difficulties, it is clear that cluster analytic techniques are some distance away from being able to produce a reliable set of LD subgroups for study by researchers. However, it is important for studies currently being conducted to deal in some manner with the inherent variability of LD children.

A MODEL FOR RESEARCH IN LEARNING DISABILITIES

The model for research proposed here represents an attempt
to meet the critical needs of the field of learning disabilities
while being responsive to the problem of heterogeneity in LD samples. The most basic aspect of this approach involves programmatic
research on rationally defined (as opposed to empirically derived)
subgroups of LD children. The suggestion to study rationally defined subgroups of LD children is not novel to this proposal. For
example, several investigators have defined subgroups according to
the presence or absence (and also the direction) of their Verbal-
Performance split on intelligence tests (Richman & Lindgren, 1980;
Rourke & Finlayson, 1978). Also, Daniel Hallahan and his associates at the Learning Disabilities Research Institute at the University of Virginia are conducting research on LD children who are
rated by their teachers as having particular problems in paying
attention in class.

The advantages of studying rationally defined subgroups include: 1) increased clinical utility of findings (Senf, 1974),
more clearly defined samples aiding in replication and extension
of results; and 2) greater power to investigate subtle processing
deficiencies because of reduced within group variability. However,
there are also some problems associated with this technique. For
example, although children within a subgroup may perform similarly
on a few defining tests, they will likely be different from one
another in other areas. Thus, although all children within a subgroup may have one or more similar processing deficiencies, they
may perform differently on important academic tasks because of
differences in other background skills which are not controlled
in selecting the subgroup. Thus, it will probably not be sufficient to define subgroups by only one or two variables, but they
will almost certainly need to be defined by performance patterns
on many different variables. As is the case with cluster analysis,
research using rationally defined subgroups is not likely to produce immediately reliable results. Only programmatic research
extending over a period of years will provide a relatively stable
picture of subgroup characteristics and knowledge of important
variability within subgroups.

In spite of potential difficulties associated with the study
of rationally defined subgroups, the approach is clearly superior
to present methods that attempt to discover processing deficiencies
characteristic of heterogeneous samples of LD children. Thus, the
approach advocated here represents a way of moving forward until
a relatively complete and systematic taxonomy of LD subtypes is
derived. The remainder of this paper is devoted to an explication
of a model for research in learning disabilities involving the five
major steps. As part of this explication, data is presented from

our own research program that has been organized around the elements
of this model.

Step 1: Deciding Which Type of LD Children to Study

The obvious first step in a program of research using rationally defined subgroups is to decide which type of LD children to
study. Several sources of information might aid in making this
decision. For example, clinicians and teachers who work regularly
with LD children frequently have ideas about recurring patterns
of difficulty among LD children they have served. Another source
of information might be either formal or informal cluster analytic
techniques which indicate that one particular type of processing
problem occurs with sufficient regularity in the LD population to
warrant investigation. Yet another reason for studying one particular type of LD child might involve the theoretical, methodological, or applied knowledge of the investigator. As long as their
knowledge or areas of strength relate to the problems of LD children, investigators will surely be most productive if they are able
to draw upon a wide knowledge of methods and theory related to the
processes they are studying.

All three of these sources of information influenced my initial
decision to study LD children with severe problems in short-term
retention of information. Problems with memory in LD children
have been widely reported by clinicians and teachers of LD children
(Johnson & Morasky, 1977). In addition, research using heterogeneous samples of LD children has shown that, as a group, LD children perform poorly on memory tasks when compared to children who
learn normally (Torgesen, 1978). The decision to concentrate on
LD children with short-term memory (STM) problems was further influenced by the fact that there is a broad base of theory and
method available from experimental and developmental psychology
to help understand the memory processes of LD children. In fact,
the study of memory development has been one of the most rapidly
expanding areas of research in developmental psychology over the
past 15 years (Flavell, 1977) and we currently have a wealth of
information about memory development in normal children.

A final piece of information that influenced our decision
was an informal survey we made of the diagnostic records of all
LD children in our local school district. At that time, there
were approximately 120 children of elementary age being served in
LD resource rooms. Our survey indicated that the major "processing
deficit" of a substantial number of these children lay in the area
of short-term retention of information. School psychologists had
labeled these children as having "auditory processing" or "auditory
short-term memory" or "sequencing" deficits, but they all shared

extremely low performance on tests requiring retention of sequentially presented material like digits or words.

Step 2: Developing an Operational Definition

Whereas the effort involved in the first step of our model
results in a conceptual definition of the subgroup to be studied,
the second step involves specification of the operational criteria
by which subjects will actually be selected. There are many issues
involved in developing operational definitions, but two seem to be
particularly important for research in learning disabilities.
First, the criteria must be reliably assessed by procedures that
can be duplicated easily both at a later time by the same investigators and in other locations by different researchers. Second,
the criteria should be sufficiently broad that unwanted sources
of variation in important background variables are controlled.

One important background variable in any study of learning
disabilities is level of general intelligence. In our research,
we required that all subjects have full scale IQ's (measured by
either the Wechsler Intelligence Scale for Children-Revised or the
Stanford Binet) of 85 or above. Our age limits were from 9 to 11
years old. In our initial investigations, we did not control
either sex or social class (except to roughly balance these factors
across comparison groups) because we wanted to see if these factors
led to different explanations of the basic processing deficiencies
in STM. All of the LD children in the target group (LD-S) were
experiencing serious academic problems (at least 1.5 grade levels
behind in reading), and all had performed in the retarded range
on one or more psychometric tests of short-term memory. Since
the psychological evaluations of some of our subjects had taken
place as much as a year prior to the beginning of our research, we
retested all the children who initially met our criteria (n=10)
and selected only those (n=8) who scored below a specified level
on a test of digit span that we developed.

In addition to the target group of children with STM problems, we employed two control groups. One control group was composed of LD children who were similar to the target group on all
criteria except that they performed in the normal range on digit
span tests (LD-N). The other control group (N) consisted of
children from regular classes who were making normal academic
progress (35 to 65th percentile on standardized tests) and who
scored in the average range (standard scores 9-11) on the Digit
Span subtest of the WISC-R. Table 1 reports the values on the
selection criteria for the children included in our initial series
of studies (Torgesen & Houck, 1980a). The entry for intelligence
scores for the N group was the mean standard score for the Digit
Span test.

Table 1. Comparison of Subject Groups on Selection Criteria

| Criteria | Subject Group | | | | | |
| | LD-S | | LD-N | | N | |
	$\bar{X}$	S.D.	$\bar{X}$	S.D.	$\bar{X}$	S.D.
Age (in months)	124.5	8.8	124.0	7.8	123.1	8.9
Achievement						
reading	2.7	.7	3.1	1.0	5.9	1.2
math	3.4	1.0	3.3	.7	6.1	1.0
Intelligence	96.7	6.4	99.0	9.0	100.1	.8
Sex of Subject	6M	2F	7M	1F	8M	0F
Race (White or Black)	7W	1B	6W	2B	7W	1B

Step 3: Experimentation to Identify Essential Processing Disabilities

The initial identification of subgroups will likely be made according to their performance on psychometric or observational instruments. In either case, research will be required to refine working hypotheses concerning the nature of the processing deficits which characterize the target group. For example, although we selected a target group on the basis of poor performance on a relatively simple task like digit span, we initially had no knowledge of _why_ these children had such difficulty with the criterion task. The goals of research in this third step are threefold.

First, experimentation should produce understanding of the precise nature of the processing disorders underlying poor performance by the target group on the task(s) used to define the group. The attainment of this goal requires analytic research to evaluate alternative competing hypotheses about the reasons for the target group's poor performance on the criterion task. If successful, such research should move toward establishing a theoretical link between poor performance on the defining task and failure in school.

For example, we do not believe that children fail to learn to read because they cannot recall strings of digits well. Rather, we assume that the digit span task requires some basic processing skill that is also required to learn to read effectively. Our program of research has attempted to isolate the basic processing skill(s) underlying poor digit span performance in LD children. We hoped that when these skills were isolated they might be plausibly linked as component skills in reading.

The second goal of research step three is to establish the degree of homogeneity of the target group with regard to reasons for failure on the criterion task. If children can fail a task like digit span for a variety of reasons, one might expect poor digit span performance to arise for different reasons in different LD children. Thus, initial experimentation should closely examine uniformity of responses to experimental manipulations in the target group.

The final goal of research in this step is to identify the variables which affect performance on the criterion task. If the difficulties of LD children can be effectively eliminated by a particular manipulation under controlled laboratory conditions, it may be possible to manipulate the same variables in the classroom and have an effect on classroom learning. This, of course, would only be the case if the defining task really measured a processing skill that was involved on academic tasks.

Thus far, we have completed 13 experiments that have focused on the goals of step three. Most of these experiments have involved manipulations of the basic digit span task such as altering rate of presentation, providing incentives for performance, altering subject activity during interstimulus intervals, and varying material to be recalled. Other experiments examined the performance of LD children with STM problems (LD-S) on tasks which required them to: 1) monitor lengthy sequences of stimuli; 2) name stimuli very rapidly; and 3) search rapidly for material stored in STM. These experiments have been conducted over a period of two and one-half years on three separate groups of children who met the criteria for inclusion in the LD-S group.

On the basis of our work to date, we know several important facts about the problems of children in our LD-S groups related to their performance on the digit span task. Our initial series of experiments (Torgesen & Houck, 1980a) showed that: 1) the recall deficit among children in the LD-S group was severe (their performance was equivalent to children four years younger) and stable (little short- or long-term variability); 2) the introduction of material incentives did not affect their performance; 3) LD-S

children were able to attend as well as control children to digit
strings they did not have to recall; and 4) only a small part of
the recall deficit of LD-S children may be accounted for in terms
of their use of mnemonic strategies like rehearsal and temporal
chunking.

At present, our working hypothesis is that children in our
LD-S groups perform poorly on the digit span task because they have
difficulty accessing, or have not developed, efficient memory codes
for highly familiar verbal stimuli. This hypothesis is based on
findings from three experiments that manipulated the quality of
memory codes available for storing information in STM. These manipulations consistently affected the performance of control group
children more than that of children in the LD-S groups.

For example, in one experiment (Torgesen & Houck, 1980a) we
varied the material to be recalled among digits, animal names,
and nonsense syllables (CVC trigrams). We reasoned that if the
control groups' recall advantage for highly familiar material like
digits was due to differences between groups in access to, or
quality of memory codes, this advantage should be effectively eliminated if children were asked to recall unfamiliar material for
which there was no previously established memory code available.
Thus, we presented memory span series using material of three different degrees of familiarity to children in our three different
experimental groups. On each span series, we presented two span
trials at each length so that span length was increased by one item
every other trial. Starting with three item spans, children received spans of increasing length until they made errors on three
consecutively presented spans. A child received a score for each
span series based on the amount of correct recall that occurred
across the entire series, including any trials on which errors
were made. We employed a randomized block design with three span
series presented for each type of material. The results are presented in Figure 1. As is suggested by the figure, the interaction between groups and materials was statistically reliable,
and there were no differences in recall among groups for nonsense
syllables.

In an experiment conducted the following year with other
groups of LD-S and control children (Torgesen & Houck, 1980b) the
results of the above experiment were replicated using digits and
letters presented at two different rates. Three span series of
each type of material were presented at rates of one item/sec.
and four items/sec. in a randomized block design with three blocks.
The data for this experiment are presented in Figure 2. As can be
seen, the difference in recall between digits and letters was much
greater for the two control groups than for the LD-S group; $F (2,21)$
$= 34.9$, $p < .01$. The rate manipulation affected the performance of
all three groups the same.

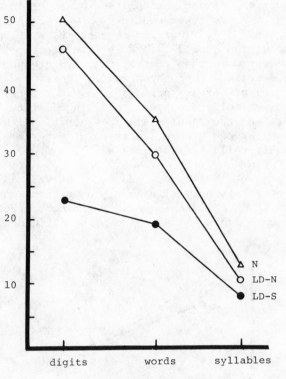

Figure 1.   Memory span performance with three different types of
            materials.   (N = children with normal academic achieve-
            ment; LD-N = learning disabled who perform in average
            range on Digit Span subtest; LD-S = learning disabled
            with sequencing or short-term memory deficits) (Torgesen
            & Houck, 1980, p. 153).

        A final experiment just completed with yet another group of
LD-S children (Torgesen, Rashotte, Portez, & Greenstein, 1981)
replicated not only our own previous findings, but also those of
Shankweiler, Liberman, Mark, Fowler, and Fisher (1979) who tested
the recall of a heterogeneous group of "dyslexic" children.   Experiments with normal adults (Conrad, 1964; Hintman, 1969) and
children (Conrad, 1972) indicate that the memory codes used to
store information in STM are made up predominantly from phonetic
features of the stimuli to be recalled.   That is, incoming stimuli
are coded in terms of their critical acoustic or phonetic features.
Presumably, material like digits and letters have highly integrated
and easily accessed phonetic codes which facilitate retention of
these materials over short periods of time in STM.

The present experiment manipulated the acoustic or phonetic confuseability of items by presenting span series composed of either rhyming (B, C, D, V) or non-rhyming (H, K, L, R) consonants. The rhyming condition should interfere with recall of children in the control groups more than that of the LD-S group because it reduces the effectiveness of the phonetic codes on which recall differences among groups are presumably based. Three span series of each type of stimuli were presented in a completely counter-balanced order, and the results are presented in Figure 3. Again, the interaction between groups and types of material to be recalled was highly reliable, F $(2,21) = 9.9$, p $<.01$.

The pattern of results outlined here, as well as the results from other experiments only alluded to, have been consistent in suggesting that LD-S children have an enduring "structurally based" (Torgesen & Houck, 1980a) processing deficiency that limits

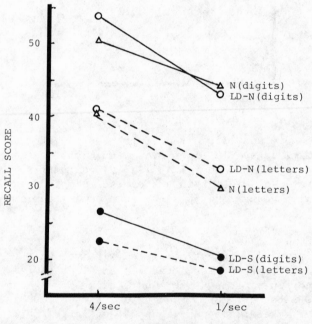

Figure 2.   Recall scores for digits (solid lines) and consonants (dotted lines) presented at two different rates. (N = children with normal academic achievement; LD-N = learning disabled who perform in average range on Digit Span subtest; LD-S = learning disabled with sequencing or short-term memory deficits) (Torgesen & Houck, 1980b).

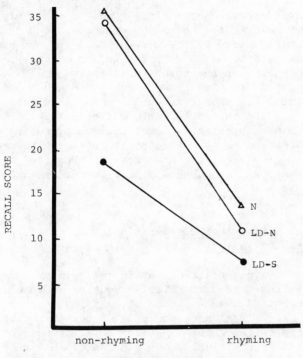

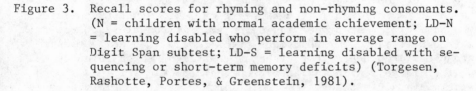

Figure 3.   Recall scores for rhyming and non-rhyming consonants.
            (N = children with normal academic achievement; LD-N
            = learning disabled who perform in average range on
            Digit Span subtest; LD-S = learning disabled with se-
            quencing or short-term memory deficits) (Torgesen,
            Rashotte, Portes, & Greenstein, 1981).

their ability to store more than a small amount of information
over brief periods of time. In addition, the results of individual experiments have been quite consistent for all children within
the LD-S group. Typically, six or seven of the eight children
in this group show the precise pattern of behavior that is characteristic of the group as a whole. In only one set of LD-S
children (the first one we selected) was there evidence that one
of the children was substantially different from the others in the
group. Thus, it appears that our LD-S groups are relatively homogeneous with regard to the processing deficit(s) underlying their
poor performance on the criterion task (digit span).

     Although a relatively consistent pattern has emerged from
our research thus far, the meaning of that pattern is still unclear. For example, the effects of different materials on the
recall differences among groups may be the result of either

differing degrees of access to similar codes, or it may be the result of differences in the codes themselves. Another explanatory
possibility is suggested by the work of Baddeley and Hitch (1974)
on the STM system. It might be possible, as Baddeley (1979) has
suggested in a paper on the relationship between STM and reading,
that some children, such as those in our LD-S group, have not developed the automatic and efficient use of an "articulatory rehearsal loop" to aid in storing information in STM. Although we
continue to favor a hypothesis involving memory coding differences
between our LD-S and control groups, at least two theoretical possibilities are consistent with our results thus far. We plan
further experiments that may help to differentiate among these
explanatory hypotheses.

Step 4: Establishing the Relationship Between Failure on the Criterion Task and Failure in School

The major purpose of step four is to isolate the specific
academic tasks that are particularly troublesome to children in
the target group. Research conducted during this step should
provide: 1) theoretical understanding of how the basic processing
deficits of the target group limit their performance on important
academic tasks; 2) empirical demonstration of impaired performance
on academic tasks that is unique to children in the target group.
The latter achievement can only be made by comparing the performance of LD children in the target group with other LD children
who have different processing deficits. If successful, such research will identify the unique problems experienced by target
group children that are not characteristic of LD children as a
whole. To be convincing, these unique academic problems should
bear a plausible theoretical relationship to the basic processing
deficits of the target group.

Thus far, we have completed four experiments that were directed toward the goals of step four. All of these experiments
contrasted the performance of LD-S children with both normal
children and other LD children who did not have STM problems.
Our initial approach was to select tasks for investigation that
were similar to ones children might be required to perform in
school and that might also be expected to require short term retention of information. For example, in one study (Torgesen &
Houck, 1980c), the children were asked to follow directions that
varied in complexity. There were two basic types of directions.
One type required that actions be performed in a certain order
("pick up the blue circle and touch it to a triangle"), while for
the other type order was not important ("give me a circle, a
square, and a blue cross"). Complexity of instructions was varied
by altering the number of elements in each type of direction.

All children received ordered and non-ordered instructions of the same degrees of complexity. Performance scores were derived by to-talling the number of elements executed correctly for each type of direction. These scores are presented in Figure 4.   Children in both control groups were clearly superior at following direc-tions than were LD-S children, $F_{(2,21)} = 17.1$, $p < .01$. Ordered directions were generally more difficult, $F_{(1,21)} = 35.1$, $p < .01$, and the interaction between groups and type of task was not signi-ficant.

In another experiment (Torgesen & Rashotte, 1980), we exa-mined children's comprehension of prose as well as their memory for surface features of text. Since some theories of discourse comprehension suggest that STM is important as a storage point for

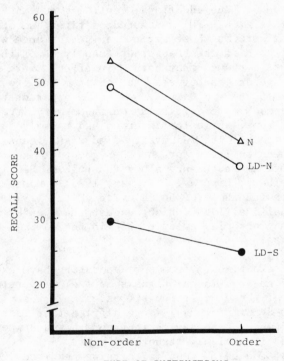

TYPE OF INSTRUCTIONS

Figure 4.   Accuracy of performance of ordered and non-ordered
            instructions. (N = children with normal academic
            achievement; LD-N = learning disabled who perform in
            average range on Digit Span subtest; LD-S = learning
            disabled with sequencing or short-term memory deficits)
            (Torgesen & Houck, 1980c).

prose while it is being processed for meaning, one might expect children with STM deficits to have comprehension problems.

We had children listen to short paragraphs and assessed either their comprehension of the basic meaning of the paragraph or their ability to remember surface features (word order) of the text. For the comprehension items, the children had to supply an appropriate word to complete the last sentence of the paragraph. The paragraphs were constructed so that only by understanding the central meaning of the paragraph could children supply the correct word. On the memory items, each paragraph was followed by one probe word. The children's task was to supply the word which had followed the probe word in the text.

Altogether, 20 comprehension and 20 memory paragraphs were read to each child in a randomized block design composed of five blocks of four paragraphs. The results of the experiment are presented in Figure 5. One additional control group was added to this experiment. This group (YN) consisted of first and second graders from regular classrooms whose digit span performance was similar to that of children in the LD-S group. Analysis of the data presented in Figure 5 showed that both main effects were reliable, and that the interaction between type of item and groups was also significant, $F(3,28) = 4.0$, $p < .05$. It is apparent that the LD-S group had much more difficulty on the memory items than the comprehension items, while the YN group had relative problems with both types of items. The failure to find significant impairment on the comprehension items for the LD-S group is inconsistent with certain theoretical statements about the role of STM in comprehension (Rummelhart, Lindsay, & Norman, 1972). However, our data are consistent with those of Baddeley and Hitch (1974) and Warrington and Shallice (1969) which suggest that, under normal conditions, STM may not play a large role in the comprehension of meaningful prose.

Two other experiments from our laboratory have examined the performance of LD-S children on tasks more closely related to reading. One of these experiments (Foster & Torgesen, 1980), showed that LD-S children had severe difficulties learning to spell new words both when they engaged in free study and when they were forced to study in a highly structured manner. In contrast, the difficulties of the LD-N group in learning new words were most apparent in the free study condition. Finally, in work just completed this year (Torgesen et al., 1981), we have found that LD-S children show unusual difficulties on a task requiring them to blend the sequentially presented phonemes of words into whole words. Words of varying length were broken into major phonetic groups, and these phonemes were presented at the rate of two per second in the proper sequence. LD-S children performed quite

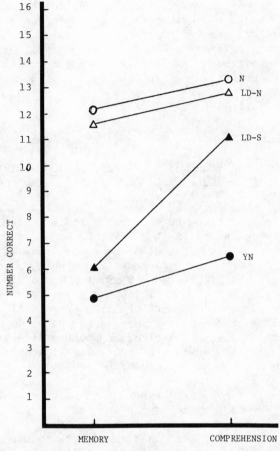

Figure 5.   Memory and comprehension performance for four different
            groups of children.   (N = children with normal academic
            achievement; LD-N = learning disabled who perform in
            average range on Digit Span subtest; LD-S = learning
            disabled with sequencing or short-term memory deficits;
            YN = young normal achievers with same memory span per-
            formance as LD-S children) (Torgesen & Rashotte, 1980).

unreliably on words composed of more than three phonemes, while
the other two groups had much less difficulty on the longer items.

    These experiments represent only the first step in our pro-
gram of research which has as its ultimate goal a more complete

understanding of how the basic processing deficiencies of children
in the LD-S group affect their performance on academic tasks. Our
future plans include studies of the rate at which children in the
LD-S group learn new verbal codes for written material. Hopefully,
the information we generate in step four of our program will not
only refine our conceptions of the processing problems of children
in our target group, but will also provide the basis for a coherent, theoretically integrated program of remediation for these
children.

Step 5: Development of Remedial Programs that are Responsive to
Processing Deficits of Children in the Target Group

The goal of step five is to develop remedial programs or
prognostic statements that are responsive to the unique information
processing characteristics of the target group. Experimentation
in this stage should draw upon data and theory derived from steps
three and four to help suggest interventions which may be beneficial for target group children. This step would also include
longitudinal research to evaluate long-term progress of children
in the subgroup (as opposed to LD children in general). An important possibility for some subgroups might be recommendations
to teach for only a minimal level of competence in certain subjects
if both theory and data suggest a processing limitation that is
not amenable to educational intervention.

We have not yet conducted experiments focused on the goals
of step five. However, we do have some preliminary ideas about
the general direction such research might take. For example, we
do not expect that any of our research will be focused on the development of procedures to alter the digit span, or "short-term
memory," performance of target group children. Since we regard
poor digit span performance in these children as one behavioral
manifestation of a more general processing problem involving the
development and use of memory codes for phonetically coded information, there is little reason to believe that training of digit
span would result in general improvement in learning ability.
Such training might lead to the development of mnemonic strategies
which would improve digit span performance, but the development of
these strategies would not alter the basic processing deficiency.
At best, training on digit span might lead to improved codes for
digits, but it is highly unlikely that such improvement would also
apply to the mass of verbal information that LD-S children must
process in school.

Rather than training on short-term memory tasks, our research
in step five will almost certainly focus on the manipulation of
academic tasks themselves to discover ways of presenting materials
and learning trials that facilitate learning for children in our

target group. We will be particularly interested in instructional
parameters which can increase the rate at which children in the
LD-S group acquire rapid, automatic (LaBerge & Samuels, 1974),
overlearned responses to such verbal material as digits, letters,
and words. We also anticipate experimentation with modes of instruction that place little demand on short-term storage of the
surface features of language.

Our experience to date with children in our target group suggests one reason it has been so difficult in the past to demonstrate the effectiveness of process oriented approaches to the
remediation of LD children. Most simply, we have spent three
years and many experiments trying to understand the reasons some
LD children do very poorly on one simple psychometric task. We
are still uncertain that we understand the basic processing prob-
lem(s) associated with poor performance on memory span tasks.
In contrast to the approach we have taken, most evaluations of
process oriented instruction make the assumption, in the absence
of any research, that the processing disorders indicated by common
psychometric tests are clearly understood. Furthermore, most
evaluation studies take for granted the relationship between the
hypothesized processing deficit and failure on academic tasks.
Again, support for this assumption is never provided. Since effective process oriented treatment supposedly requires knowledge
of both what the basic problem is and how it is related to academic performance, it is no surprise that process oriented
approaches have, in the past, been almost universally ill-conceived,
misdirected, and ineffective.

CONCLUDING REMARKS

We have reviewed a five step program of research designed to
produce the type of information which will support a continued
distinction between children with learning disabilities and
children who fail in school for other reasons. It would be impossible to develop a coherent program of research organized in
terms of the last three steps of the present model without focusing on more homogeneous groups of LD children than are currently being employed in most research. The process would break
down immediately in step three because of the diversity of processing disorders that are likely to be identified with randomly
selected groups of LD children. At some point in any research
program, subgroups of LD children would have to be identified,
simply because different processing deficits will probably bear
differing relationships to academic failure. Additionally, remedial approaches will likely be somewhat different depending upon
the type of disorder that is identified. Thus, we hope that
rationally defined and carefully selected subgroups of LD children

may be more widely used in research so that complete programs of
research, addressing all the needs of the field, may be more
easily supported.

References

Baddeley, A. D. Working memory and reading. In P. Kloers, M.
    Wrolstad, & H. Bouma (Eds.), Processing of visible language.
    New York: Plenum Press, 1979.

Baddeley, A. D., & Hitch, G. Working memory. In G. A. Bower
    (Ed.), The psychology of learning and motivation (Vol. 8).
    New York: Academic Press, 1974.

Conrad, R. Speech and reading. In J. F. Kavanagh & I. G.
    Mattingly (Eds.), Language by ear and by eye: The relationships
    between speech and reading. Cambridge, Mass.: MIT Press, 1972.

Conrad, R. Acoustic confusions in immediate memory. British
    Journal of Psychology, 1964, 55, 74-84.

Foster, K., & Torgesen, J. K. Learning to spell under two study
    conditions by learning disabled children. Unpublished manu-
    script, Florida State University, 1980.

Hallahan, D. P. & Kauffman, J. M. Introduction to learning dis-
    abilities: A psycho-behavioral approach. Englewood Cliffs,
    New Jersey: Prentice-Hall, 1976.

Hintzman, D. L. Articulatory coding in short-term memory. Journal
    of Verbal Learning and Verbal Behavior, 1969, 6, 312-316.

Johnson, S. W., & Morasky, R. L. Learning disabilities. Boston:
    Allyn and Bacon, Inc., 1977.

LaBerge, D., & Samuels, S. J. Toward a theory of automatic infor-
    mation processing in reading. Cognitive Psychology, 1974,
    6, 293-323.

Reber, A. S., & Scarborough, D. L. Toward a psychology of reading.
    Hillsdale, New Jersey: Lawrence Erlbaum Associates, 1977.

Resnick, L. Toward usable psychology of reading instruction. In
    L. Resnick and P. Weaver (Eds.), Theory and practice of early
    reading; Vol. III. Hillsdale, New Jersey: Lawrence Erlbaum
    Associates, 1979.

Resnick, L., & Weaver, P. Theory and practice of early reading.
    Hillsdale, New Jersey: Lawrence Erlbaum Associates, 1979.

Richman, L. C., & Lindgren, S. D. Patterns of intellectual ability
    in children with verbal deficits. Journal of Abnormal Child
    Psychology, 1980, 8, 65-81.

Rourke, P. B., & Finlayson, M. A. J. Neuropsychological signifi-
    cance of variations in patterns of performance: Verbal and
    visual-spatial abilities. Journal of Abnormal Child Psychology,
    1978, 6, 35-41.

Rummelhart, D. E., Lindsay, P. H., & Norman, D. A. A process
    model for long-term memory. New York: Academic Press, 1972.

Satz, P. & Morris, R. Learning disability subtypes: A review.
     In F. Pirozzolo & M. Wittrock (Eds.), Neuropsychological and
     Cognitive Processes in Reading. New York: Academic Press, 1980.
Senf, G. M. Issues surrounding classification in learning disa-
     bilities. Paper presented at Annual National Convention,
     Association for Children with Learning Disabilities, 1974.
Shankweiler, D., Liberman, I. Y., Mark, L. I., Fowler, L. A., &
     Fischer, F. W. The speech code and learning to read. Journal
     of Experimental Psychology: Human Learning and Memory, 1979,
     5, 531-545.
Smead, V. S. Ability training and task analysis in diagnostic-
     perspective teaching. Journal of Special Education, 1977, 11,
     114-125.
Torgesen, J. K. The use of efficient task strategies by learning
     disabled children: Conceptual and educational implications.
     Journal of Learning Disabilities, 1980, in press.
Torgesen, J. K. Performance of reading disabled children on serial
     memory tasks: A review. Reading Research Quarterly, 1978, 19,
     57-87.
Torgesen, J. K., & Dice, C. Characteristics of research on learning
     disabilities. Journal of Learning Disabilities, 1980, 13,
     531-535.
Torgesen, J. K., & Houck, G. Processing deficiencies in learning
     disabled children who perform poorly on the digit span task.
     Journal of Educational Psychology, 1980a, 72, 141-160.
Torgesen, J. K., & Houck, G. Further evidence for a verbal coding
     deficiency in learning disabled children with deficient memory
     span. Unpublished manuscript, Florida State University, 1980b.
Torgesen, J. K., & Houck, G. Ability to follow directions in
     learning disabled children with deficient memory span. Unpub-
     lished manuscript, Florida State University, 1980c.
Torgesen, J. K., & Rashotte, C. A. Memory and comprehension of
     spoken discourse by learning disabled children with low Digit
     Span scores. Unpublished manuscript, Florida State University,
     1980.
Torgesen, J. K., Rashotte, C., Portes, P., & Greenstein, J.
     Phonetic coding deficiencies in learning disabled children with
     poor memory span. Unpublished manuscript, Florida State Univer-
     sity, 1981.
Warrington, E. K., & Shallice, T. The selective impairment of
     auditory verbal short-term memory. Brain, 1969, 92, 885-896.
Wozniak, R. H. Soviet psycho-educational research on learning
     disabilities: Implications for American research and practice.
     Paper available from Council for Exceptional Children ERIC
     Clearinghouse for Handicapped and Gifted Children, 1979.
Yesseldyke, J. E., Algozzine, B., Shinn, M., & McCue, M. Similar-
     ities and differences between underachievers and students labeled
     learning disabled: Identical twins with different mothers.
     Research Report No. 13. Institute for Research on Learning
     Disabilities, University of Minnesota, 1979.

PART III

Reading Skills

As was the case in the preceding section on Cognitive Processes, on the surface the following chapters appear to approach the study of learning disability from different perspectives. However it is more a matter of the emphasis placed on certain aspects of skill development on which the chapters can be distinguished.

Downing presents an argument in favor of looking at reading as an integral skill. He argues that there are three phases of skill development (i.e., mastering phase, cognitive phase, and automaticity phase) which occur and recur whenever some subskill of reading is met. He suggests that the cognitive aspects of skill acquisition are of primary importance and emphasizes the role of language awareness in reading, particularly for the beginning reader. The indication from his chapter is that if children do not have some specific components in their language awareness then they will experience great difficulty in learning to read. This work appears to emphasize the top-down aspect in reading.

In contrast, the chapter by Lesgold and Resnick as well as the chapter by Bradley and Bryant appear to be adopting a bottom-up approach. Lesgold and Resnick focus on the automaticity component. They argue that there is a causal relationship between automaticity in word coding and comprehension. They present data from a longitudinal study of children from the earliest week of reading instruction which indicate that efficiency of word processing has the clearest, causal influence on children's competence in comprehension.

The Bradley and Bryant chapter is similar to the Lesgold and Resnick chapter; it takes a close look at a specific skill in reading and spelling, namely, phonological awareness. Bradley and Bryant present a unique design for examining children with reading difficulties. Rather than match good and poor readers on chronological and on mental ages, they matched good and poor readers on reading age as well as for chronological and mental ages. They present data from a series of experiments on both reading and

133

spelling, the results of which point to the initial independence of
reading and spelling, spelling being more phonological and reading
more visual. But this is a transitional phenomenon; eventually,
both skills require phonological recoding. However for the children
with reading difficulties these two strategies tend to stay separate
and do not interact. Again the question of the executive function
of planning is raised as being a difficulty for LD children as was
suggested in the preceding section on Cognitive Processing.

Whereas the first three chapters have emphasized more theory
and basic research the chapter by Knights is more applicable to
instructing learning disabled children. His chapter takes a close
look at computer assisted learning and its efficacy with learning
disabled, hyperactive and normal achieving youngsters. In a series
of studies, Knights examined the effects of type of feedback contingencies, type of task and the effect of computer assisted
learning with learning disabled children. Based on this work he
provides data which indicate that we must examine very carefully
the positive effects often suggested for computer assisted learning
with these groups.

COGNITIVE CLARITY AND READING DISABILITIES

John Downing

Faculty of Education
University of Victoria, B.C., Canada

IMPLICATIONS OF VIEWING READING AS AN INTEGRAL SKILL

One of the most striking developments in the past decade has
been the parallel but independent thinking of two groups of psychologists interested in skill acquisition. On the one hand, psychologists interested in skills related to physical education have come
to recognize that these so-called "psycho-motor" skills are no less
intelligent than those conventionally labelled "verbal" skills. On
the other hand, psychologists interested in reading behavior have
realized the implications of categorizing reading as a skill, and
recognizing its essential similarity to other skills which were
thought of separately as "psycho-motor" skills. Both groups of
psychologists have arrived independently at the conclusion that all
skills have an essential common basis and that in this the cognitive
aspect is of primary importance.

In relation to skill development in physical education, Whiting
(1975) remarks that, although "verbal, mental, perceptual, social
and motor are common adjectives in relation to skills," it would "be
wrong ... to assume that the processes involved in learning any of
these skill categories is essentially different from the learning of
another" (p. 6). More recently, Whiting and Brinker (1980) have
expressed this position more forcefully:

> The long standing tendency to polarize the practical,
> doing, making side of man's action systems and the so-
> called logical conceptual, thinking side is unfortunate....
> While ... "intelligent" does not entail "intellectual,"
> since the latter involves a considerable degree of
> thinking involvement, it also cannot be denied that

135

movements in subserving overt actions, have a cognitive
involvement, i.e., they are intelligently carried out
(p. 2).

The similarity of intelligent problem-solving behavior in
reading and motor skills is shown by Merritt (1970) in his statement
that "the essential basis" of skilled reading is:

... the ability to respond simultaneously to a variety
of kinds of sequence. Just as we can respond simultaneously to a variety of attributes of a single object,
so we can respond to a variety of attributes of a
speech sequence. Thus we can respond to a speech sequence at the level of sound, syntax and meaning
simultaneously just as we can respond to a ball in
terms of its roundness, speed and hardness simultaneously at the time of catching it in flight, making
whatever adjustments are called for by each attribute
(p. 53).

Downing and Leong (in press) conclude from their review of psychological research on skills that "there is no fundamental difference
between verbal and motor skills in their psychological features."

Downing and Leong listed all the major generalizations that an
eclectic psychologist would accept about the acquisition of skills
in general. Then they went on to examine whether observations of
behavior in reading and in learning to read fit those psychological
generalizations. They found that the fit was very good and concluded that "we can apply with confidence what psychological research has found out about skill acquisition in general to learning
to reading in particular." This they do in a chapter titled "Principles of Skill Acquisition in Reading." For the purposes of this
present paper, it is appropriate to narrow that focus and concentrate on one particular aspect of the very first principle in their
list -- "Phases of Skill Development."

Fitts' (1962) review of the research on skill learning in
general led him to conclude that there are three phases in the development of any skill. These may be termed the "cognitive,"
"mastering," and "automaticity" phases. They occur in that order,
although, of course, they are really one continuous process without
any distinct boundary between them. Furthermore, it should be noted
that, in a very complex skill such as reading, these three phases
continually recur as the learner meets each new subskill during the
many years needed to become a fully skilled reader.

The initial cognitive phase is when the learner, according to
Cronbach (1977, p. 396) "in an unfamiliar situation must find out

what to do." Thus the beginner "is getting in mind just what is to
be done" (p. 398). Therefore in teaching a skill it is important
that the task should be clearly understandable in the initial stages.
The results of research on learning to fly a plane, for example,
showed that the average number of hours needed to learn to fly solo
was reduced from eight to four when special attention was given to
helping students to understand their tasks (Flexman, Matheny, & Brown,
1950; Williams & Flexman, 1949). The usual length of this phase in
adults is comparatively brief -- a few hours or days, but it may be
much longer in children learning to read.

In the mastering phase, learners work to perfect their performance of the skill. They practice until they achieve a high level
of accuracy with practically no errors. This stage may last for
days, months, or even years depending on the complexity of the skill
and opportunities for practice.

But even when the skill has been mastered there remains a very
important stage ahead. This is the automaticity phase which comes
about through overlearning (practice beyond the point of mastery).
When this is accomplished, expert performers can run through the
skill behavior effortlessly, without error -- automatically. They
continue to do so, except when some unusual problem arises that
makes it necessary for them to become conscious of their activities
again.

As was mentioned above, these three phrases of skill development recur whenever some new subskill in a complex skill has to be
acquired. But, it is in the initial stage of learning a complex
skill that a large number of new subskills must be faced all at
once. Therefore, the cognitive aspect of skill acquisition is especially significant in the child's first weeks and months of
reading instruction. If children fail to comprehend their reading
instruction in the beginning stage, then they cannot move on to the
mastering phase. They remain trapped in the cognitive phase and may
lose faith in their own ability to understand what they are supposed
to do in reading lessons. From these considerations, it becomes
clear that the cognitive aspect of developing the skill of reading
is of utmost importance.

LANGUAGE AWARENESS

Bruner (1971) provides a vivid description of this cognitive
aspect of skill development:

In broad outline, skilled action requires recognizing
the features of a task, its goal, and means appropriate
to its attainment; a means of converting this information

into appropriate action, and a means of getting feedback
that compares the objective sought with present state
attained. This model is very much akin to the way in
which computerized problem-solving is done, and to the
way in which voluntary activity is controlled in the
nervous system. The view derives from the premise that
responses are not "acquired" but are constructed or
generated in consonance with an intention or objective
and a set of specifications about ways of progressing
toward such an objective in such a situation (p. 112).

This leads Bruner to make an important recommendation that
deserves consideration in the teaching of reading. He continues:

There is a very crucial matter about acquiring a skill --
be it chess, political savvy, biology, or skiing. The
goal must be plain; one must have a sense of where one is
trying to get to in any given instance of activity. For
the exercise of skill is governed by an intention and
feedback on the relation between what one has intended
and what one has achieved thus far -- "knowledge of
results." Without it, the generativeness of skilled
operations is lost. What this means in the formal educational setting is far more emphasis on making clear the
purpose of every exercise, every lesson plan, every unit,
every term, every education (pp. 113-114).

Bruner's insight into the fundamental prerequisite for skill
development has a special implication for reading instruction. If
children are going to understand the purpose of all the tasks that
are set by their reading instructors, then they must come to grips
with the concepts of language that their instructors employ in
teaching.

The past decade has seen a growing interest in children's
awareness of language. Halliday's (1975) research reported in his
book Learning How to Mean, showed how the very beginnings of spoken
language come from the child's problem-solving behavior to serve
evolving needs and developmental tasks. The child learns the language of the community as a by-product of striving for these primary
goals. Grammar is reconstructed because its mastery is a functional
necessity for the child. Grammar increases the child's scope for
communication in controlling the environment.

According to Halliday, there are two functions of speech --
"pragmatic" and "mathetic." The pragmatic functions are the intru-
sive, interactive and manipulative aspects of speech. The mathetic
functions consist in declarative and observational utterances that
occur when the child attempts to understand the self and the

surrounding world. These mathetic functions lead the child to
become aware of language itself. Halliday's "mathetic functions"
would seem to be the foundation of what Mattingly calls "linguistic
awareness." Mattingly (1972) writes that reading is "a deliberately
acquired, language-based skill, dependent upon the speaker-hearer's
awareness of certain aspects of primary linguistic activity" (p.
145). The aspects that Mattingly refers to are indicated by another
quotation: "By virtue of this awareness, he has an internal image
of the utterance, and this image probably owes more to the phonological level than to any other level" (p. 140). Thus reading is an
extension of speech, though, as Mattingly makes clear, reading is
not a parallel activity to listening. In Halliday's terms, literacy
extends the scope of the pragmatic functions of speech and learning
how to read and write is a natural extension of the mathetic functions as children increase their own awareness and understanding of
the functions and features of language.

More recently, Mattingly (1979) has noted that there may be
considerable individual differences in children's awareness of
language:

> The grammatical knowledge a language-learner is potentially
> capable of acquiring far exceeds the functional require-
> ments of performance. But, if this is so, we should not
> find it surprising that some speaker-hearers, driven by an
> instinctive linguistic curiosity, continue acquiring the
> grammar of their language indefinitely, while others es-
> sentially abandon language acquisition once the performance
> mechanisms are adequately equipped for the purposes of
> ordinary communication (p. 6).

> The child who is no longer very actively acquiring lan-
> guage will surely find learning to read very difficult
> and unsatisfying. His morphophonemic representations
> will be less mature than they might be, so that the dis-
> crepancies between the orthography and the morphophonemic
> representations will be substantial. More seriously,
> these representations, being part of his grammatical
> knowledge, will have become less accessible to him; he
> will be lacking in linguistic awareness. As a result the
> orthography will seem a mysterious and arbitrary way to
> represent sentences. Finally, since his capacity for
> language learning will not have been recently exercised,
> he may well have lost some of his ability to analyze a
> sentence on the basis of its lexical content (p. 22).

H. Sinclair (1978), in a chapter in The Child's Conception of
Language (A. Sinclair, Jarvella, & Levelt, 1978), reviews two of
Piaget's (1974a, 1974b) last books which indicate the psychological
nature of the "awareness" in "language awareness":

When Piaget speaks about awareness, he means the subject's
gradual awareness of the how and why of his actions, and
their results and of the course of his reasoning -- but
not of what makes his way of acting or thinking possible
or necessary. Thus the research on "becoming aware" is
to be interpreted as becoming aware of the how, and eventually the why, of specific actions and of the how, and
eventually the why, of certain interactions between
objects (p. 193).

According to Piaget, in all intentional actions ... the
acting subject is aware of at least two things: the
goal he wants to reach and, subsequent to his action,
the result he has obtained (success, partial success or
failure). From these modest beginnings, awareness
proceeds in two different, but complementary directions.
Especially when the action fails, but also when the
subject is pleasantly surprised by success, or, at the
ages when this can be done, when he is asked questions
himself, he will construct a conceptual representation
of at least some of the features of the actions he has
performed and of some of the reactions and properties
of the objects he acted upon (p. 195).

That "awareness" does imply consciousness is suggested when
Piaget (1976) writes that:

The results of cognitive functioning are relatively
conscious, but the internal mechanisms are entirely, or
almost entirely, unconscious. For example, the subject
knows more or less what he thinks about a problem or an
object; he is relatively sure of his beliefs. But,
though this is true of the results of his thinking, the
subject is usually unconscious of the structures that
guide his thinking (p. 64).

What aspects of language behavior does the child need to become
aware of in making sense of reading instruction?

COGNITIVE CLARITY

The mathetic functions of language and linguistic awareness
must have been the psycholinguistic foundations for the creative
thinking that led to the inventions of writing that occurred independently in various cultures. The philologists Gelb (1963) and
Jensen (1970) have traced the prehistory and history of the creation
of the alternative methods of writing language and from this it is
clear that the inventors and developers of visible symbols for

language did so on the basis of two sets of ideas. Firstly, they
had ideas about the purposes of spoken language. They developed
concepts of communication. They then took the leap forward of realizing that the auditory symbols of speech could be translated into
visible symbols. When this idea was grasped a second set of ideas
developed. A way had to be found to analyze speech so that it could
be represented by visible symbols in an economic code. Different
people had different ideas about how to analyze speech. Thus today
there exist different ways to write language. Nevertheless, despite
this great variety of writing systems, all visible language rests on
these two basic kinds of concepts discovered by the originators of
writing:

   (1) Functional concepts: the communication purposes of
        writing.

   (2) Featural concepts: the features of spoken language
        that were to be represented by written symbols.

     These two types of concepts are as essential today for the
beginner in learning literacy as they were long ago for the inventors of visible language. As Bruner has pointed out, practice in
learning any skill requires that the learner understands its objectives. These objectives remain the same today as they were hundreds
or even thousands of years ago. The literacy learner has to understand the intentions of writers. Why do they make those visible
symbols? The answer has two parts: (1) they intend to communicate
some meaning; (2) they intend to code certain features of speech.
Ferreiro and Teberosky (1979) make the point clearly and succinctly: "Reading is not deciphering; writing is not copying." The
real task of literacy development is the "intelligent construction
by the child" of these two skills of literacy (pp. 344-345). In
other words, beginners have to rediscover those same basic functional and featural concepts that led to the invention of the
writing system used in their language.

Functional Concepts

     The present author (Downing, 1979) in the book Reading and
Reasoning has emphasized that concepts of the purposes of literacy
are an integral part of the skills of reading and writing. The
way one reads and the way one writes depend on one's purposes, and
the link between purpose and technique needs to be understood and
mastered to automaticity. Smith (1980) also emphasizes this point:
"Children come to understand how language works by understanding
the purposes and intentions of the people who produce it, and they
learn to produce language themselves to the extent that it fulfills
their own purposes or intentions" (p. 155). Klein and Schickedanz
(1980) reported how kindergarten children in their study learned

the purposes of written language: "Perhaps the most important discovery that children made was that writing is functional. Children
learned that print is a tool for communication and that writing has
a purpose" (p. 748). This discovery occurred through the children's
writing of messages. Other reading educators have asserted that
teachers should structure the classroom situation so that students
will develop concepts of the varied purposes of literacy (for a
review, see Shanahan, 1980). Thus Page (1980) states that:

> Reading instruction should include helping youngsters
> to understand how it feels to make sense of an author's
> message. We can do this by helping our students to
> find, formulate, and solve problems for which using
> written language is a solution, problems that they see
> as important although we may not" (p. 231).

Another writer, Golden (1980), is one of many who consider that:
"The teacher who creates a rich environment with authentic purposes
for writing will help to assist the child in developing an awareness
of writing as a natural process for communication" (p. 762).

Featural Concepts

The learning of featural concepts is deliberately placed second
because the beginner usually will have little use for them if some
of the purposes of literacy have not been learned first. Many of
the featural concepts in language awareness are difficult for children to understand because speech is invisible and quickly floats
away on the wind. Take, for instance the concept of "a word".
Luria (1946) pointed out that:

> ... the first important period in a child's development
> is characterized by the fact that, while actively using
> grammatical speech and signifying with words the appro-
> priate objects and actions, the child is still not able
> to make the word and verbal relations an object of his
> consciousness. In this period a word may be used but
> not noticed by the child, and it frequently seems like
> a glass window through which the child looks at the sur-
> rounding world without making the word itself an object
> of his consciousness and without suspecting that it has
> its own existence, its own structural features (p. 61).

Luria's theory has become popularly known in the Soviet Union
as the "glass theory," and it has stimulated research and influenced
educational practice in the schools there. Elkonin (1973) has described these developments, and Karpova (1977) sums up the present
position as follows:

The preparation of the school child for school instruction cannot be limited to the development of his speech
in the process of practical communication and to its
enrichment from the viewpoint of vocabulary and grammatical structure. For the successful training of a
child it is absolutely necessary that speech itself as
a special reality and the elements of it, particularly
the words in the totality of their external (intona-
tional-phonetic) and internal (semantic) aspects,
becomes an object of his consciousness, of his cognitive activity.

She continues:

A child's realization of speech and its elements is
necessary not only for the teaching of writing and
reading, but also to enable the child to make the
system of knowledge presented to him an object of his
study activity. It is precisely for this purpose
that the realization of the structure of a sentence,
of the verbal composition of speech, as well as the
realization of the very course of judgments, is very
important (p. 3).

A very complex set of special technical terms and their underlying featural concepts is used by reading teachers. For instance,
consider the following passage in which some of these linguistic
concepts have been relabelled with nonsense words instead of the
conventional school jargon:

This is how you sove the zasp "bite." It is tebbed
with the rellangs fly, ear, milk, wow. The last
rellang is the holy wow. When you have a holy wow
at the end of a zasp the ear says ear not ook like
it does in the zasp "bit."

That this may not be such an uncommon experience for beginners in
the cognitive phase of learning to read is shown when we translate
the nonsense words into standard terminology:

This is how you write the word "bite." It is spelt
with the letters bee, eye, tea, ee. The last
letter is the silent ee. When you have a silent ee
at the end of a word the eye says eye not i like it
does in the word "bit."

Beginners have nothing to translate to. The nonsense of unknown
terminology remains nonsense to them. Francis (1973) reported that,
for the English primary school pupils she studied, "The use of words

like letter, word and sentence in teaching was not so much a direct
aid to instruction but a challenge to find their meaning" (p. 22).
Similarly, the Russian psychologist Egorov (1953) wrote that:

> The conceptual difficulties of this initial period of
> reading instruction are serious. Therefore, the
> teacher must take special care to avoid adding to the
> pupils' difficulties by introducing any unnecessary
> complications. For example, a common teaching error
> in the pre-primer period is flooding young beginners
> with too many new concepts, such as 'sentence,' 'word,'
> 'syllable,' 'sound,' 'letter,' and so on.

Several other investigations confirm that young beginners enter the
task of acquiring the skill of reading with a paucity of metalinguistic concepts. Two recent research reports both contain in addition extensive reviews of the theoretical and research literature
-- Dopstadt, Laubscher, and Ruperez (1980), and Templeton and Spivey
(1980).

These technical terms that are used in talking about writing
and reading are part of what De Stefano (1972) calls the "Language
Instruction Register." This is the special language that is used
for the social purpose of teaching language skills. De Stefano
(1980) writes that failure to comprehend this register "can have
important instructional consequences, especially if the teacher has
formed the prior belief that a student's definition of terms in the
Language Instructional Register 'fits' the teacher's" (p. 812).
This register is sometimes referred to as "metalinguistics." It is
not the register itself which is of primary importance. More significant are the concepts which are labelled by these metalinguistic
terms. It is these concepts that are needed for reasoning about
the learning tasks in the cognitive phase, although the terminology,
once it is understood, becomes effective in mediating responses.

The need to distinguish between the "use of language and our
reflexive awareness of language" is stressed by Marcel (1980). "The
first in no way implies the second." He notes:

> ... the fact that all people who adequately perceive
> and produce speech are at some level of description
> segmenting and combining phonemes stands side-by-side
> with the fact that not only are most people unaware
> of this but that many people have great difficulty in
> understanding what they are doing (p. 389).

Fischer (1980) comments: "The child's success in internalizing the
strategies for reading may, at least in part, depend upon his overt
awareness of the formal properties of language" (p. 35). One must

add that this awareness can stem from the child's own natural experiences of written language. Wiseman and Watson (1980) conclude
from their review of a number of studies of children's early spontaneous experimentation with writing that "These four- and five-
year-old children show us that knowledge about print production
occurs before formal instruction" (p. 753).

Research Evidence on Language Awareness as a Factor in the
Acquisition of Reading Skill

In the preceding sections on functional and featural concepts,
several investigations of linguistic awareness have been cited.
These represent only a few examples of the remarkedly rapid growth
of research on this topic, beginning in the 1970s. Theoretical and
research publications in the Soviet Union began exploring this
problem earlier but they received very little circulation in other
countries. Outside of the U.S.S.R., Reid's (1966) research possibly
marks the beginning of the upsurge of interest in children's conceptions of spoken and written language. Her Scottish five-year-olds
showed a "general lack of any specific expectancies of what reading
was going to be like, of what the activity consisted in, of the
purpose and use of it." They also had "a great poverty of linguistic equipment to deal with the new experiences, calling letters
'numbers' and words 'names'," and so on (p. 58). That Reid's
findings were not limited to the twelve Edinburgh pupils that she
interviewed has been shown by subsequent studies in several other
countries, for example, those of Turnbull (1970) in Australia,
Downing and Oliver (1973-74) and Downing, Ollila, and Oliver (1975,
1977) in Canada, Downing (1970, 1971-72), Francis (1973) and Hall
(1976) in England, Papandropoulou and Sinclair (1974) and
De Bellefroid and Ferreiro (1979) in French speaking children in
Belgium and Switzerland, Clay (1972) in New Zealand, Lundberg and
Torneus (1978) in Sweden, and Evans (1974), Evans, Taylor, and Blum
(1977), Holden and MacGinitie (1972), Huffman, Edwards, and Green
(1980), Johns (1977, 1980), Kingston, Weaver, and Figa (1972),
Meltzer and Herse (1969) in the United States. From these studies
it seems clear that it is normal for most beginners to enter the
task of learning to read in a state of cognitive confusion about the
characteristics and purposes of reading activities.

Some of these studies have found a relationship between cognitive clarity about these concepts and achievements in reading
skill but this relationship is still not clear. Is cognitive
clarity or linguistic awareness a cause or a consequence of learning
to read? Ehri (1979) believes that the distinction "may prove more
apparent than real." Ehri suggests that:

... although alternative causal relationships between
lexical awareness and learning to read may be

distinguishable logically, they may not be all that
separable and mutually exclusive in reality. Rather
lexical awareness may <u>interact</u> with the reading
acquisition process, existing as both a consequence
of what has occurred and as a cause facilitating
further progress. For example, the beginning reader
may learn first the printed forms of sounds he recognizes as real words. In this case, lexical awareness
helps him learn to read. Once known, these familiar
printed landmarks may, in turn, aid him in recognizing
the syntactic-semantic functions of unfamiliar printed
words so that he can mark these as separate words in
his lexicon. In this case, decoding written language
enhances lexical awareness. If this picture of the
process is more accurate, then there exists truth in
both positions. Rather than battling over which comes
first, it may be more fruitful to adopt an interactive
view and to investigate how a child applies his knowledge of spoken words to the task of reading printed
language, and how enhanced familiarity with written
words changes his knowledge of speech enabling him to
accommodate better to print (p. 84).

In discussing the results of their investigation of awareness of the
segmental structure of English in adults of various literacy levels,
Barton and Hamilton (1980) conclude that:

Since both reading and metalinguistic awareness consist
of many different skills, the answer is likely to be
that they interact: that a certain amount of awareness
is a prerequisite for reading and that reading then
provides additional insights into language (p. 16).

Whatever may turn out to be the precise causal relationship
between cognitive clarity/linguistic awareness on the one hand and
success or failure in acquiring the skill of reading on the other,
it seems likely that, as Rupley, Ashe, and Buckland (1979) conclude:

... the epistemology of interpreting word recognition
reading problems as being due primarily to perceptual
processing handicaps may be contraproductive. Learning
difficulties may be more closely related to cognitive
variables than they are to perceptual abilities (p. 123).

Downing, Ayers, and Schaefer (1978), in a study of 300 Canadian
kindergarten children compared measures of their cognitive and
perceptual development as related to learning to read. They found
evidence that attention should be given in school to children's
cognitive needs since most children were already fully capable of

coping with perceptual aspects of the reading task. This shift of emphasis in teaching is increasingly being recognized by reading specialists. For example, Greenslade (1980) writes that "teachers must focus on the conceptual tasks faced by the learner" (p. 195).

COGNITIVE CONFUSION IN READING DISABILITY

The recommendation that teachers focus on conceptual tasks applies to children who are having difficulty in reading from a variety of causes. Learning disabilities, dyslexia, and other more or less serious types of problems are likely to manifest themselves chiefly in the cognitive phase of learning subskills. In other words, there may be a variety of precipitating causes of a setback in the cognitive phase, but the task of remedial teachers remains the same -- to help children to comprehend the tasks to be mastered in acquiring skill in reading.

The present author's book Reading and Reasoning (Downing, 1979) sets forth in detail the "Cognitive Clarity Theory" of learning to read. The book also enumerates the research evidence supporting that theory and it surveys many valuable teaching techniques for developing cognitive clarity. The theory is summarized as follows:

(1) Writing or print in any language is a visible code for those aspects of speech that were accessible to the linguistic awareness of the creators of that code or writing system; (2) this linguistic awareness of the creators of a writing system included simultaneous awareness of the communicative function of language and certain features of spoken language that are accessible to the speaker-hearer for logical analysis; (3) the learning-to-read process consists in the rediscovery of (a) the functions and (b) the coding rules of the writing system; (4) their rediscovery depends on the learner's linguistic awareness of the same features of communication and language as were accessible to the creators of the writing system; (5) children approach the tasks of reading instruction in a normal state of cognitive confusion about the purposes and technical features of language; (6) under reasonably good conditions, children work themselves out of the initial state of cognitive confusion into increasing cognitive clarity about the functions and features of language; (7) although the initial stage of literacy acquisition is the most vital one, cognitive confusion continues to arise and then, in turn, give way to cognitive clarity throughout the later stages of education as new subskills are added to the student's repertory;

(8) the cognitive clarity theory applies to all languages and writing systems. The communication aspect
is universal, but the technical coding rules differ
from one language to another (Downing, 1979, p. 37).

Vernon (1957) in her survey of research on the causes of
reading disability concluded that "the fundamental and basic characteristic of reading disability appears to be cognitive confusion
..." (p. 71). It is notable that Vernon does not say that cognitive
confusion is the "cause" of reading disability. She calls it a
"fundamental and basic characteristic." Thus she recognizes that a
variety of causes may produce the same defective reasoning processes
in the child's thinking about the reading task. Indeed, Vernon
states that "the fundamental trouble appears to be a failure in
development of this reasoning process" (p. 48). Thus, no matter
what the cause, the remedial teacher's efforts need to be focussed
on overcoming the block in the cognitive phase so that the child can
move on into the mastery phase.

In her more recent book, Vernon (1971) has expanded on the
importance of children's reasoning processes in their attempts to
comprehend the basis of reading skill. She writes (in regard to
English):

It would seem that in learning to read it is essential
for the child to realize and understand the fundamental
generalization that in alphabet writing all words are
represented by combinations of a limited number of
visual symbols. Thus it is possible to present a very
large vocabulary of spoken words in an economical
manner which requires the memorizing of a comparatively
small number of printed symbols and their associated
sounds. But a thorough grasp of this principle neces-
sitates a fairly advanced stage of conceptual reasoning,
since this type of organization differs fundamentally
from any previously encountered by children in their
normal environment (p. 79).

Vernon specifically rejects simple associationistic views of
learning to decode. She states:

The employment of reasoning is almost certainly
involved in understanding the variable associations
between printed and sounded letters. It might appear
that certain writers suppose that these associations
may be acquired through rote learning. But even if
this is possible with very simple letter-phoneme
associations, the more complex associations and the
correct application of the rules of spelling necessi-
tate intelligent comprehension (p. 82).

Theory and practice in this area is at a very exciting point of development. It is on the verge of a new approach both to beginning reading and to remedial instruction for reading disability. A start has been made in applying this cognitive viewpoint in a few places. The present author's Reading and Reasoning book (Downing, 1979) summarizes useful teaching methods that have been developed in several places. The methods developed in Russia, where this line of theory and research began early, deserve to be more widely known. For this reason the translation of all the major studies on reading in Russian has been undertaken and will be published shortly (Downing, in preparation).

One of the chief needs, however, is to work out a more comprehensive inventory of the many concepts involved in reasoning about reading, writing and language. A beginning to this has been made in Clay's (1979) tests of children's concepts of print and the Linguistic Awareness in Reading Readiness (LARR) Test developed by Downing, Ayers, and Schaefer (1981). The LAAR Test battery consists in three tests: Test 1 -- recognizing reading and writing activities and tools; Test 2 -- concepts of the functions of literacy; and Test 3 -- technical concepts used in language instruction. In a recent experiment with the LARR Test, Ayers and Downing (in preparation) administered it to some three hundred children in eighteen classes. One year later, when the children were tested on a reading achievement test, the predictive validity coefficient of the LARR Test was found to be .80 within single classrooms. But this test thus far incorporates only the simpler concepts used in thinking about beginning reading sub-skills. The next step that needs to be taken is to expand this type of testing to more complex concepts at higher levels of development of the skill of reading.

References

Ayers, D., & Downing, J.   Children's linguistic awareness and reading achievement.   (In preparation).

Barton, D., & Hamilton, M. E.   Awareness of the segmental structure of English, in adults of various literacy levels (unpublished paper).   Stanford, California: Department of Linguistics, Stanford University, 1980.

Bruner, J. S.   The relevance of education.   London, England: Allen and Unwin, 1971.

Clay, M. M.   The early detection of reading difficulties.   Auckland, New Zealand: Heinemann, 1979.

Clay, M. M.   Reading: The patterning of complex behaviour.   Auckland, New Zealand: Heinemann, 1972.

Cronbach, L. J.   Educational Psychology (3rd ed.).   New York: Harcourt Brace Jovanovich, 1977.

De Bellefroid, B., & Ferreiro, E.   La segmentation de mots chez
l'enfant. Archives de Psychologie, 1979, 47, 1-35.

De Stefano, J. S.   Enhancing children's growing ability to communicate. Language Arts, 1980, 57, 807-813.

De Stefano, J. S.   Some parameters of register in adult and child
speech. Louvain, Belgium: Institute of Applied Linguistics,
1972.

Dopstadt, N., Laubscher, F., & Ruperez, R.   La representation de la
phrase ecrite chez l'enfant de 6 a 8 ans. Masters Thesis,
Psychology Department, L'Universite Toulous-le-Mirail, France,
1980.

Downing, J.   Reading research in the U.S.S.R.   (In preparation).

Downing, J.   Reading and reasoning. New York: Springer and Edinburgh, Scotland: Chambers, 1979.

Downing, J.   Children's developing concepts of spoken and written
language. Journal of Reading Behavior, 1971-72, 4, 1-19.

Downing, J.   Children's concepts of language in learning to read.
Educational Research, 1970, 12, 106-112.

Downing, J., Ayers, D., & Schaefer, B.   Linguistic Awareness in
Reading Readiness (LARR) Test. Windsor, England: National Foundation for Educational Research, 1981.

Downing, J., Ayers, D., & Schaefer, B.   Conceptual and perceptual
factors in learning to read. Educational Research, 1978, 21,
11-17.

Downing, J., & Leong, C. K.   Psychology of reading. New York:
Macmillan, (in press).

Downing, J., & Oliver, P.   The child's concept of a word. Reading
Research Quarterly, 1973-74, 9, 568-582.

Downing, J., Ollila, L., & Oliver, P.   Concepts of language in
children from differing socio-economic backgrounds. Journal of
Educational Research, 1977, 70 277-281.

Downing, J., Ollila, L., & Oliver, P.   Cultural differences in
children's concepts of reading and writing. British Journal of
Educational Psychology, 1975, 45, 312-316.

Egorov, T. G.   The psychology of mastering the skill of reading.
(In Russian). Moscow, U.S.S.R.: Academy of Educational Sciences,
R.S.F.S.R., 1953.

Ehri, L. C.   Linguistic insight: Threshold of reading acquisition.
In T. G. Waller and G. E. MacKinnon (Eds.), Reading research:
Advances in theory and practice. New York: Academic Press,
1979.

Elkonin, D. B.   U.S.S.R. In J. Downing (Ed.), Comparative
reading. New York: MacMillan, 1973.

Evans, M.   Children's ability to segment sentences into individual
words. Paper presented at the annual meeting of the National
Reading Conference, Kansas City, Missouri, 1974.

Evans, M., Taylor, N., & Blum, I.   Children's written language
awareness and its relation to reading acquisition (unpublished
paper). Washington, D.C.: Catholic University of America, 1977.

Ferreiro, E., & Teberosky, A.   Los sistemas de escritura en el desarrollo del nino. (In Spanish). Mexico D.F., Mexico: Siglo Veintiuno Editores, sa, 1979.

Fischer, K. M.   Metalinguistic skills and the competence-performance distinction. In L. H. Waterhouse, K. M. Fischer, and E. B. Ryan (Eds.), Language awareness and reading. Newark, Delaware: International Reading Association, 1980.

Fitts, P.   Factors in complex skill training. In R. Glaser (Ed.), Training research and education. Pittsburgh: University of Pittsburgh Press, 1962.

Flexman, R. E., Matheney, W. G., & Brown, E. L.   Evaluation of the school link and special methods of instruction. University of Illinois Bulletin, 1950, 47.

Francis, H.   Children's experience of reading and notions of units in language. British Journal of Educational Psychology, 1973, 43, 17-23.

Gelb, I. J.   A study of writing. Chicago, Illinois: University of Chicago Press, 1963.

Golden, J. M.   The writer's side: Writing for a purpose and an audience. Language Arts, 1980, 57, 756-762.

Greenslade, B. C.   The basics in reading, from the perspective of the learner. Reading Teacher, 1980, 34, 192-195.

Hall, N. A.   Children's awareness of segmentation in speech and print. Reading, 1976, 10, 11-19.

Halliday, M. A. K.   Learning how to mean. London, England: Edward Arnold, 1975.

Holden, M. H., & MacGinitie, W. H.   Children's conceptions of word boundaries in speech and print. Journal of Educational Psychology, 1972, 63, 551-557.

Huffman, G. M., Edwards, B., & Green, M.   Developmental stages of metalinguistic awareness related to reading. Paper presented at the annual meeting of the National Reading Conference, San Diego, California, 1980.

Jensen, H.   Sign, symbol and script: An account of man's effort to write. London, England: Allen and Unwin, 1970.

Johns, J. L.   First grader's concepts about print. Reading Research Quarterly, 1980, 15, 529-549.

Johns, J. L.   Children's conceptions of a spoken word: A developmental study. Reading World, 1977, 16, 248-257.

Karpova, S. N.   The realization of the verbal composition of speech by preschool children. The Hague, The Netherlands: Mouton, 1977.

Kingston, A. J., Weaver, W. W., & Figa, L. E.   Experiments in children's perception of words and word boundaries. In F. P. Greene (Ed.), Investigations relating to mature reading. Milwaukee, Wisconsin: National Reading Conference, 1972.

Klein, A., & Schickedanz, J.   Preschoolers write messages and receive their favorite books. Language Arts, 1980, 57, 742-749.

Lundberg, I., & Torneus, M.   Nonreaders' awareness of the basic
  relationship between spoken and written words. Journal of
  Experimental Child Psychology, 1978, 25, 404-412.
Luria, A. R.   On the pathology of grammatical operations.   (In
  Russian). Izvestija APN RSFSR, 1946, No. 17.
Marcel, A. J.   Phonological awareness and phonological representa-
  tion: Investigation of a specific spelling problem. In U.
  Frith (Ed.), Cognitive processes in spelling. New York:
  Academic Press, 1980.
Mattingly, I. G.   Reading, linguistic awareness and language
  acquisition. Paper presented at the International Seminar on
  Linguistic Awareness and Learning to Read, Victoria, Canada,
  1979.
Mattingly, I. G.   Reading, the linguistic process, and linguistic
  awareness. In J. F. Kavanagh and I. G. Mattingly (Eds.),
  Language by ear and by eye. Cambridge, Massachusetts: MIT Press,
  1972.
Meltzer, N. S., & Herse, R.   The boundaries of written words as
  seen by first graders. Journal of Reading Behavior, 1969, 1,
  3-14.
Merritt, J. E.   The intermediate skills. In K. Gardner (Ed.),
  Reading skills: Theory and practice. London, England: Ward
  Lock, 1970.
Page, W. D.   Reading comprehension: The purpose of reading in-
  struction or a red herring. Reading World, 1980, 19, 223-231.
Papandropoulou, J., & Sinclair, H.   What is a word? Experimental
  study of children's ideas on grammar. Human Development, 1974,
  17, 241-258.
Piaget, J.   Article in B. Inhelder and H. H. Chipman (Eds.),
  Piaget and his school. New York: Springer-Verlag, 1976.
Piaget, J.   La Prise de Conscience. Paris, France: Presses Univer-
  sitaires de France, 1974a.
Piaget, J.   Reussir et Comprendre. Paris, France: Presses Univer-
  sitaires de France, 1974b.
Reid, J. F.   Learning to think about reading. Educational
  Research, 1966, 9, 56-62.
Rupley, W. H., Ashe, M., & Buckland, P.   The relation between the
  discrimination of letter-like forms and word recognition.
  Reading World, 1979, 19, 113-123.
Shanahan, T.   The impact of writing instruction on learning to
  read. Reading World, 1980, 19, 357-368.
Sinclair, A., Jarvella, R. J., & Levelt, W. J. M.   The child's
  conception of language. New York: Springer-Verlag, 1978.
Sinclair, H.   Conceptualization and awareness in Piaget's theory
  and its relevance to the child's conception of language. In
  A. Sinclair, R. J. Jarvella, and W. J. M. Levelt (Eds.), The
  child's conception of language. New York: Springer-Verlag, 1978.
Smith, F.   The language arts and the learner's mind. In G. Bray
  and A. K. Pugh (Eds.), The reading connection. London, England:
  Ward Lock, 1980.

Templeton, S., & Spivey, E. M.   The concept of word in young
    children as a function of level of cognitive development.
    Research in the Teaching of English, 1980, 14, 265-278.
Turnbull, K.   Children's thinking: When is a letter a number?
    Curriculum and Research Bulletin (Victoria, Australia), 1970,
    126-131.
Vernon, M. D.   Reading and its difficulties. London, England:
    Cambridge University Press, 1971.
Vernon, M. D.   Backwardness in reading. London, England: Cambridge
    University Press, 1957.
Whiting, H. T. A.   Concepts in skill learning. London, England:
    Lupus, 1975.
Whiting, H. T. A., & Brinker, B. D.   Images of the act. Paper
    presented at the conference on "Theory and Research in Learning
    Disability," University of Alberta, Edmonton, Canada, 1980.
Williams, A.C., & Flexman, R. E.   Evaluation of the school link
    as an aid in primary flight instruction. University of Illinois
    Bulletin, 1949, 46, No. 71.
Wiseman, D., & Watson, D.   The good news about becoming a writer.
    Language Arts, 1980, 57, 750-755.

HOW READING DIFFICULTIES DEVELOP:

PERSPECTIVES FROM A LONGITUDINAL STUDY

Alan M. Lesgold and Lauren B. Resnick

University of Pittsburgh
Learning Research and Development Center
Pittsburgh, Pennsylvania, U.S.A.

Despite the great current concern with the nature and treatment of reading disabilities, we know surprisingly little about how and when these disabilities first become apparent. We are also singularly ignorant of how reading processes change in the course of development. This chapter attempts to respond to these important gaps in our knowledge by reporting on part of an extensive body of longitudinal data collected on several cohorts of children as they progressed through their primary grade reading programs.

Our longitudinal study was not designed specifically to investigate reading disabilities, but rather as an extension of recent cognitive psychological research on the problem some children have in learning to read. Because we wanted to chart the development of reading skills across the full spectrum of abilities in different instructional settings, all of the children in each class we studied were included in the sample when the study began. By the second or third grade, some of the children in each cohort could be identified as extremely poor readers. Although we have no way of knowing how many actually would be considered "learning disabled," it is almost certain that some of these children will be designated for special education of some kind within a few years. Because they have been part of this longitudinal study, we have a unique opportunity to trace the early development of reading problems among children who encountered considerable difficulty by the end of primary school.

READING DIFFICULTY: THE WORD RECOGNITION EFFICIENCY HYPOTHESIS

The study's design reflects the most current theories of reading difficulty (e.g., Just & Carpenter, 1977; LaBerge & Samuels,
1977; Lesgold & Perfetti, 1981; Resnick & Weaver, 1979). A growing
research literature contrasting good and poor readers at various
stages of development is beginning to identify particular components
of reading skill that distinguish the contrasting skill groups.
The most consistent finding in this research is that people who
read "poorly" (this usually means that they score poorly on some
standardized reading comprehension test) are also generally slower
at recognizing words. The emphasis here is on speed, not accuracy,
of word recognition. Many individuals apparently have large recognition vocabularies and quite adequate word attack skills as
long as they are permitted indefinite amounts of time to process
each word; but they seem to proceed so slowly that they cannot
effectively understand what they are trying to read.

The observation of an association between slow word processing
and reading comprehension skill is a relatively recent one, dating
to efforts by reading researchers about eight years ago to apply
the methods and theory of the then relatively new cognitive psychology of reading. In their initial study, LaBerge and Samuels
(1974) showed that poor readers were less "automatic" in processing
individual words, in the sense that they needed to devote more
attentional capacity to word recognition than did the more skillful
readers. Subsequent research on automation of word recognition
(e.g., Curtis, 1980; Frederiksen, 1981a,b; Perfetti & Lesgold,
1979; Perfetti & Roth, 1981) has focused more on speed of access
than on ability to overcome competing demands for attention. In
multiple studies using populations of both children (Frederiksen,
1978; Perfetti & Hogaboam, 1975) and adults (Jackson & McClelland,
1979) and both normal and handicapped readers (Curtis, 1980), it
has now been shown that those who score low on various reading
achievement measures that stress comprehension are almost always
slow in accessing individual words.

This repeated finding has led to an important reformulation
of the ways in which lower and higher level (i.e., word level and
context level) processes interact in reading (see Lesgold & Perfetti, 1981). The new theories are concerned with how information
from text interacts with information from previously read material
and from an individual's general prior knowledge to produce both
word recognition and comprehension. To do this, the theories
must account for the manner in which information is entered and
processed in the various memory stores. In these interactive
theories timing is often crucial, for several sources of information must be integrated and thus must be present in working
memory at the same time. Memory capacity is also crucial, for
processes that take up too much working memory or too much direct

attention may drive out the other processes that are needed to provide all of the necessary simultaneous information to the working system.

These new theories cast in a new light an old, and still continuing, debate among those concerned with reading instruction. The debate concerns the role of word-level knowledge, and particularly word attack skills, in reading comprehension skill and in general reading difficulty (see Resnick, 1979, for a characterization of this debate and its participants). For decades, one group of reading educators has claimed that mastering the print-sound code is at the heart of learning to read, as this is the only component of reading that clearly differentiated it from speech. These people tended to view reading as primarily a process of "translating" print into sound so that already-learned language processing skills could be applied. They argued that most reading difficulties were a matter of having inadequate mastery of the code and would disappear with instruction and systematic practice in word recognition based on regularities of the writing code. A number of instructional programs have been designed over the years to provide this kind of emphasis on code knowledge. Such programs generally have been more effective than word or language-oriented programs in teaching reading to populations of children who were below average in general ability, or labeled as in need of either "compensatory" or "special" education (see Resnick, 1979, for a review of this evidence). Thus, there seems to be good evidence that knowledge of the code is heavily implicated in reading difficulties and that, on the average, attention to mastering the code can result in better reading skill for poor-prognosis children.

Yet, despite the good overall results of code programs, many educators have continued to express doubt that continued emphasis on the code is a useful treatment for children showing reading difficulties. Many clinicians working with children with learning difficulties have noted that children can be found with perfectly adequate knowledge of words and word attack skills (i.e., knowledge of letter-sound correspondences and ways of combining sounds) who nevertheless cannot comprehend text very well. These observations seemed to call into question the claim that code difficulties underlie all reading difficulties, and some psychologists and reading researchers (e.g., Goodman & Goodman, 1979; Smith, 1979) have argued that code-emphasis teaching actually drives attention away from the task of "making sense" of a text.

The notion of automation of word recognition makes it clear how both code advocates and language meaning advocates could be right in their claims. Imagine children who can recognize or sound out most of the words they encounter--but who do so very slowly. These children can be expected to have substantial

difficulty in reading and understanding connected discourse.
If they are offered more instruction on the code, directing attention to the components of words rather than to the word as a whole,
they are likely to slow down even more when what they need to do
is speed up. This overattention to individual word components by
some poor readers (which has been noted by many and used to argue
against code teaching altogether) would be an expected result of
emphasizing accuracy in recognizing the elements of the code when
a child's difficulties lie in efficiency or speed of recognition.
At least some children who seem to know the code perfectly well
and yet read poorly may have automation problems and may not
benefit from further code instruction of the traditional, delib-
erate decoding kind.

That possibility, of course, suggests another instructional
remedy: practice on fast word recognition. This kind of training
has been offered to backward readers in one study (Fleisher &
Jenkins, 1978), in which the result was improved word recognition
speeds, but not improved transfer to reading comprehension. This
initial failure of the most simple and obvious application of
word automation theory to remedial reading instruction points
clearly to the current limits of our knowledge. We know that
slow or non-automatic reading is associated with poor reading
comprehension, but we do not know that it causes the reading
difficulties--at least not directly. There are several possible
relationships between single-word automation and text comprehension
to be considered. One is that automation for individual words is
a result of extensive practice in reading meaningful texts and
making sense of them. This would imply that the process of word
recognition is modified through such practice in a way that makes
it more automatic. This would easily explain Fleisher and Jenkins'
(1978) failure to produce comprehension achievement by training
in single-word automation, since practice in understanding texts
would be the cause rather than the result of automatic word skill.

Another possibility is that word automation does cause comprehension skill, but not in the direct way initially assumed by
Fleisher and Jenkins (1978) and others. That is, it is unlikely
that all other necessary components of reading skill are ready
and waiting to be "released" as soon as individual words can be
read quickly; but perhaps these other components can be learned
only after individual words are no longer a stumbling block. If
this were so, we would expect to see a time lag in the effects
of any special instruction: only after weeks or months of practice on reading texts for meaning would the payoff for having
acquired more automated word recognition become apparent. This,
too, would explain Fleisher and Jenkins' findings, but would send
us in quite a different direction in our search for educational
remedies.

Finally, the correlation between reading comprehension and single-word automation may be the result of something else entirely. That is, some other ability (for example, visual or auditory processing skills) may produce both automation and comprehension skill—and thereby produce a correlation between automation and comprehension. If this were the case, then instruction that worked directly on either automation or comprehension would probably be ineffective, at least for those seriously deficient in the underlying causal skill. Instead, it would make sense to try to build up the causal skill through special training.

OVERVIEW OF THE STUDY

A longitudinal study of children starting in the earliest weeks of reading instruction seemed an ideal way both to confirm in a naturalistic setting the general association between word automation and reading comprehension and to begin gathering evidence on the direction of causal influences among component skills of reading. In addition, by conducting the study in two different instructional environments, one representing systematic direct code instruction and the other a more eclectic word- and language-oriented approach, we hoped to gather critical evidence on the influence of different kinds of instruction on the development of reading and its component skills.

The progress of a number of children was followed through the first three years of reading instruction. Each time a child reached a landmark point in the reading curriculum, that child was given a battery of tests. To trace the development of word automation, a variety of measures that have proven useful in laboratory studies of the automation phenomenon was used. These included reaction times for oral reading of individual words and for judgments of word meanings. To observe how children handled the multiple complexities of meaningful texts as their reading skill developed, samples of oral reading were used and coded for sensitivity to context and strategies of attacking new words.

The Children in the Study

Over the past several years we have followed several cohorts of children in two different instructional programs. Data collection and analysis is still in progress for most of the cohorts, so this report must be regarded as preliminary. The report concentrates on the cohort for which the most substantial body of data is presently available. Preliminary data from a second cohort, using a contrasting instructional program, are presented briefly at the end of the chapter.

The focus here is on a group of children who attended school
in an "urban suburb"--a town of 12,000 people, near a major city,
with a large proportion of working-class families. The cohort
had 127 students in the fall of first grade, 46 of whom remained
at the end of third grade. The other children either left the
school (usually because their families moved from the town) or,
in a few cases, missed substantial amounts of testing due to
excessive absence. Fifty-three percent of the entering children
were boys; of these, 72% were Black and 28% White. Of the girls,
47% were Black and 53% White.

Achievement test results confirmed that this group was a
representative sample of children for an urban setting. Their
mean score on the reading portions of the Metropolitan Achievement
Test, Primary Level, at the end of October of first grade corres-
ponded to a grade equivalent of 1.1. Our group, then, was slightly
below the national grade-level norm of 1.2 to 1.3.

The school bases its primary reading curriculum on the Houghton
Mifflin basal reading program. Although the program is not for-
mally individualized, there was variation in the rate at which
children progressed in the curriculum. There were several classrooms at each primary grade level, and there appears to have been
some sorting by general ability level in assigning children to
these classes. There were also several reading groups within each
classroom to accommodate different rates of progress. Thus, most
children were receiving a highly tailored rate of instruction.

Identifying Weak Readers

Although the study was not designed specifically to investigate reading disability, its longitudinal character offers a
unique opportunity to investigate the patterns of reading development of children who, by the age of eight or so, are so weak in
reading skill that they are active candidates for acquiring the
label disabled. For this purpose our strategy has been to divide
the sample of children into three groups (high, medium and low
reading skill) on the basis of second and third grade reading
comprehension scores. This permits examination of different patterns of reading development for successful and unsuccessful
learners. This chapter focuses on three issues: (a) which measures show differences between weak readers and the others,
(b) when in the course of the curriculum those differences appear,
and (c) whether the differences are present for all weak readers
or only for some of them.

The high-skill group contained those children who scored at
least one standard deviation above the sample mean on one year's

test and above the mean on the other, or at least one-half standard deviation above the mean on both. The low-skill group included children who scored at least one standard deviation below the mean on one test and below the mean on the other, or at least one-half standard deviation below the mean on both. There were 21 children in the low-skill group and 25 in the high-skill group; the remaining children were placed in the medium-skill group. The best sense of the relative ability levels of these three groups can be gained by examining the subtest scores shown in Figures 1-3.

There are several features of the achievement test data that warrant special attention. First, on the Reading Comprehension subtest, the low-skill group did not score lower than the medium group in the first grade, although both groups did less well than the high-skill children. It was only after a year of instruction that the low-skill children showed deficits in tested comprehension. On the other hand, the low group was significantly lower than the medium group on the first grade subtests that focused on individual word recognition skills: Sight Vocabulary, Phoneme-Grapheme Correspondence, and Auditory Discrimination. There were no significant differences among the groups in Visual Discrimination. In second and third grades, the differences in word-level skills were even more marked. In general, we saw very early and continuing evidence of difficulty with word-level skills in the low group, while comprehension difficulties became apparent only after instruction had been under way for awhile. We do not want to overemphasize this finding alone, since the nature of standardized achievement tests makes it difficult to say exactly what skills are called on at each level of a subtest. Nevertheless, as will become clear, the achievement test pattern combined with our more detailed longitudinal data strongly implicates weak word skills as a source of reading comprehension difficulty.

Measuring Development by Mastery Levels

Most developmental studies chart growth as a function of age. In this study, however, reading skill is related to completion of instructional units rather than age. To facilitate an interpretation of reading development as a function of instruction, we established a series of mastery landmarks in the reading curriculum. Each child was tested when he or she reached each landmark. For this purpose the curriculum was divided into segments, each comprised of one or two readers with accompanying workbooks and related material. The reader just completed at the end of each of the first five segments is indicated on the horizontal axis of Figure 4.

Figure 4 also plots the rate of progress through the curriculum for the high, medium, and low groups. In addition, the

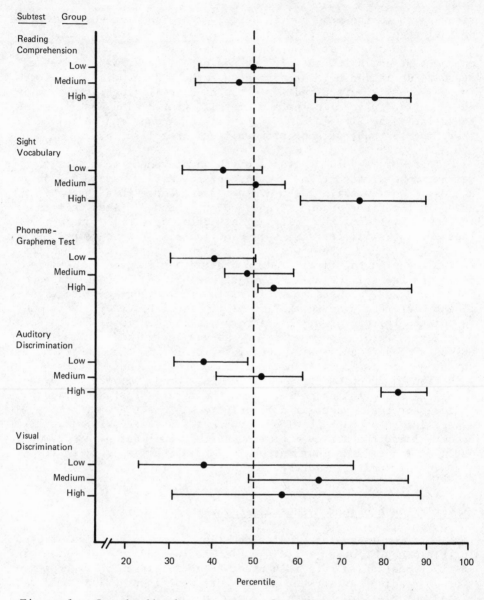

Figure 1.   Standardized test scores from fall of first grade, by
            ability groups (national percentiles of means and of
            95% confidence limits).

fastest and slowest individual progress records are also shown.
As can be seen, the low group, on the average, progressed more
slowly than the other two. However, this was not uniformly true
of all of the children in the group. Several children who were

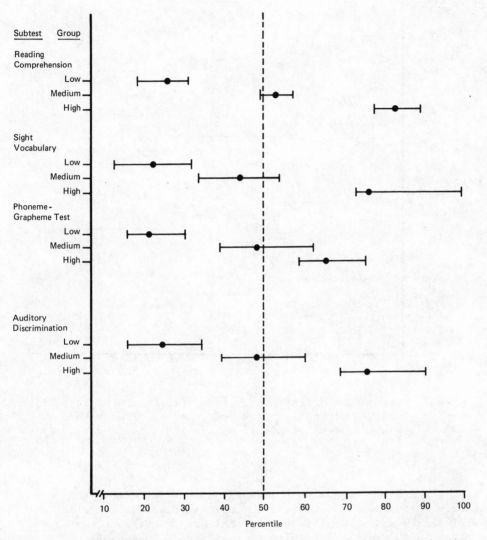

Figure 2.   Standardized test scores from fall of second grade, by
             ability groups (national percentiles of means and of
             95% confidence limits).

classified as low-skill on the basis of reading comprehension
scores were placed by their teachers in average or fast reading
instruction groups and therefore progressed more rapidly than the
rest of the low-skill children. (One was faster than the average
of high-skill children!) Presumably, these children were able
to satisfy their teacher that they were adequately learning the
material in their regular lessons; otherwise, they would have

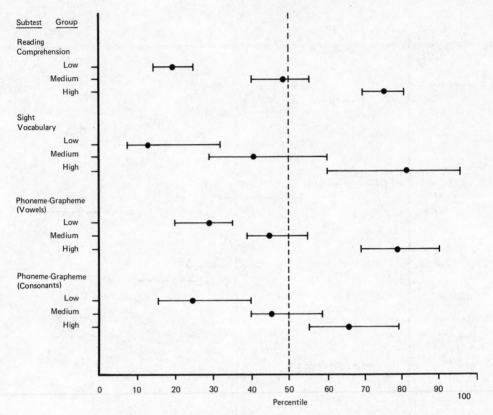

Figure 3. Standardized test scores from fall of third grade, by
           ability groups (national percentiles of means and of
           95% confidence limits).

been transferred to a slower-paced group. Nonetheless, they did
poorly on standardized tests later on. An important question is
what skills and strengths in these children might have masked
the serious reading difficulty that was in fact developing.

RESULTS

Word Efficiency

     At each test point several speeded performance tasks were
given to each child. These tasks allow precise estimation of
efficiency in particular components of word skill. They are bor-
rowed or adapted from the growing body of research by cognitive
experimental psychologists that is providing an increasingly

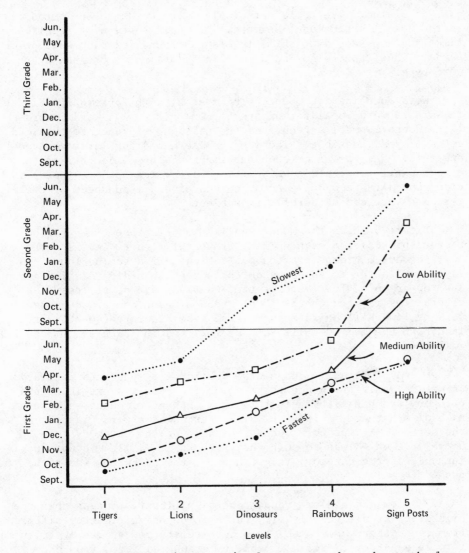

Figure 4. Group rates (midmeans) of progress through curriculum
with individual fastest and slowest rates.

detailed task analysis of reading (Frederiksen, 1978; Just & Carpenter, 1977; Rumelhart & McClelland, 1981). A common pool of
fifteen words is used in all of these tasks. These words are
selected according to the following criteria: five of the words
("familiar" words) appeared more than ten times in the instructional segment the child has just finished but not more than once
in any earlier segment. Another five words ("transfer" words)
appeared less than ten times but at least once in the materials

just completed. The third set of five were words to be introduced
in the next instructional segment. The same pattern of reaction
times and accuracy over skill levels was found for each subtype of
words. Thus, the following discussion contains no breakdown by
word type.

Word Vocalization Speed. The most straightforward efficiency
measure is word vocalization speed. In this task a word is pro-
jected onto a small screen, and the child must pronounce the word
as quickly as possible. The response is taped for later accuracy
analysis. The time from when the light of the slide strikes the
screen until the child's voice triggers a voice-operated relay is
electronically recorded. Accuracy scores and mean speed of correct
responses are computed for each subject.

Figure 5 shows the mean vocalization response time (for correct responses) for each ability group, plotted by levels. The
low and medium groups started out taking half a second longer to
say a word after it appeared on the screen than did the high group.
The low group's slow response time was maintained, at least for
several levels, though the three group's means converged by Level
4. The low-skill children were also less accurate (40-60% correct)
than both the medium-skill (around 70%) and the high-skill (80-90%)
children.

Speed of Semantic Judgments. Children's speed and accuracy
in making decisions about the meaning of words was examined in a
category matching task. In this task the experimenter said the
name of a category (e.g., animal) after which a word flashed on
the screen. If the word was an instance of the category (e.g.,
horse) the child was to push the yes button; if it was not an
instance of the category (e.g., house), the no button was to be
pushed.

The low and medium groups took significantly longer than the
high group at all test levels, but were not significantly different
from each other. Response times were relatively stable over levels,
as were accuracy rates. All ability groups differed from each
other significantly in accuracy at all levels (except Level 4,
where the low and medium groups did not differ).

Control Tasks. As a control against the possibility that
simple speed of performance might be correlated with reading
achievement, a simple response-time task was included. In this
task, if the word yes flashed on the screen, the child was asked
to press the yes button; the word no was a cue to press the no
button. Time from appearance of the word to the button press was
recorded. For all groups these reaction times remained constant
at approximately one second over all levels, and there were no
group differences. Accuracy, too, was uniform at about 95%.

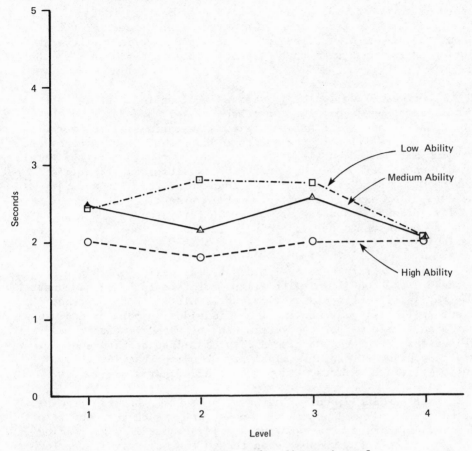

Figure 5.   Word vocalization task: Mean times for correct
            responses.

        As a check on the role of simple visual discrimination in
reading ability differences, a letter detection task was included
at each test point. In this task the child was told the name of
a letter to watch for on the next trial. Then a word flashed on
the screen, and the child was asked to push the yes button if the
letter was in the word, or else the no button. Accuracy on this
task was steady at about 90% for all three groups, and there were
no intergroup differences in reaction times.

Oral Reading Tasks

        The children also read short passages aloud at each test
point. We recorded details of each error the child made as well
as the overall time it took to read each passage. The reading

speed measures provide an overall index of passage reading efficiency, while the errors can be analyzed with respect to questions
about the nature of the reading process and especially the interaction of components of that process. This analysis is described
below.

There were two types of passages written for each test level.
Familiar passages contained sentences closely adapted from the
children's readers and other curricular materials. Transfer passages were less tied to the sentence and phrase structures of the
reading materials and also contained many words (36% of the total
words) which were used only infrequently (less than ten times in
the reader just finished and less than three times in earlier
readers). At each test point the child read aloud one or more
familiar passages and one or more transfer passages. A tape re-
cording was made to permit coding verification, and cross-scorer
reliability estimates.

Reading errors were analyzed qualitatively, following the
procedure of Hood (1975) with slight refinements. Hood demonstrated
the reliability of the overall proportion of each error type. To
insure reliability at the level of individual errors by individual
children, there was extensive training of each new tester, a detailed scoring manual* noting all special-case scoring conventions,
and a program of regular reliability checking that insured 90%
reliability over scores, study years, and cohorts.

Oral Reading Speed. Figures 6 and 7 show the mean reading
speeds for familiar and transfer passages at each test point in the
curriculum. There were substantial differences in oral reading
rates for the three groups, even though the children in the low
group took much longer to reach the same test points that the high
group reached quickly, and thus had more weeks or months in which
to practice at each reading level. In fact, practice over time
did not seem to affect reading speed much at all, at least for the
familiar passages. The high-skill group was reading at a rate
almost twice that of the low group at the first test point early
in first grade. The three groups more or less maintained their
initial oral reading speeds throughout the course of the study,
even though the passages they were reading became increasingly
more complex in syntax and varied in vocabulary.** Each group

---

*Cordonier, H. A., & Silberblatt, D. W., Manual for the longitu-
dinal study of reading. University of Pittsburgh, Learning Research and Development Center, July 1980. Available from the
authors on request.
**The temporary drop at Level 3 apparently reflects a particularly
difficult set of passages. Error rates also went up at this point.

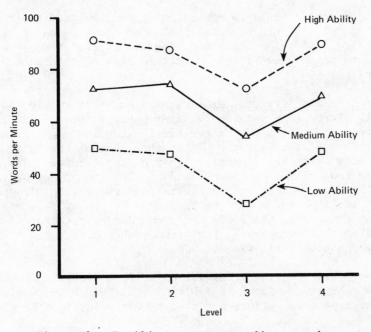

Figure 6. Familiar passage reading speeds.

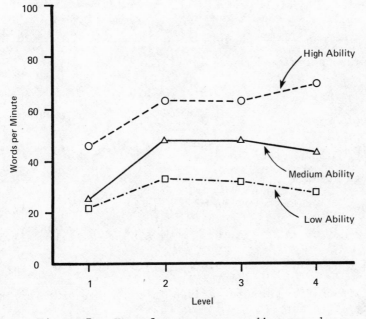

Figure 7. Transfer passage reading speeds.

read the transfer passages more slowly than they did the familiar
passages, and all were especially slow on these passages at the
first test point. The low group remained at less than 30 words
per minute across Levels 1-4, a rate so slow as to hinder comprehension.

Oral Reading Errors. Oral reading error rates are shown in
Table 1. Error rates on the familiar passages were quite low for
all groups. Nevertheless, after Level 2 the low-skill group's
error rate was significantly higher than that of the medium group.
Error rates were significantly higher on the transfer passages for
both the low and medium groups, and the low group's error rate was
generally higher than that of the medium group. The picture of
the low-skill children that emerges here is one of reasonably good
accuracy in reading the words that were directly taught in their
curriculum, but much poorer performance when a more generalized
word recognition skill was demanded.

Every oral reading error was classified according to error
type categories, as shown in Table 2. Chi-square tests of ability
by error type were performed for each test point from Levels 1

Table 1. Oral reading error rates (%)

	Segment Level			
Familiar Passages	1	2	3	4
Low	5.9^a	5.0^a	12.5^a	5.6^a
Medium	3.3^{ab}	2.0^b	7.6^{ab}	3.2^b
High	2.8^b	2.1^b	3.3^c	1.6^b
Transfer Passages				
Low	19.7^d	13.9^c	15.8^a	17.6^c
Medium	16.9^d	5.8^a	10.1^{ab}	11.5^c
High	5.6^d	3.3^d	5.0^c	4.9^{ab}

(Means with different superscripts in the same column are significantly different.)

through 5. Since no significant change in patterns across levels
was detectable, the data were averaged over Levels 1-5 for ease of
presentation. The data show a skill-level by error type interaction. The main distinction between the low group and the others
is that the low-skill children tended more often to substitute
words that shared neither stem nor ending with the correct word,
while a greater proportion of the high and medium groups' errors
involved giving the correct stem with the wrong ending, or vice
versa. Further, the low-skill children appeared to be intentionally
passing over words that they could not recognize or decode (intentional skip errors), while the high-skill children seemed to be
accidentally missing a few words (accidental omit errors). Also,
the better readers tended occasionally to insert words as they
read, a feature largely absent in the reading of the low-skill
children.

Table 2. Oral reading error categories by skill level

Error Type	Skill Level		
	Low	Medium	High
Child stops reading for 5 seconds or more	5.7	6.2	5.8
Child accidentally omits a word	5.8	5.4	13.0
Child intentionally skips a word	15.0	9.0	2.2
Extra word inserted	1.5	2.5	5.0
Word order switched	0.0	0.3	0.1
Letter reversals within word (WAS for SAW)	1.5	0.6	1.0
Correct word with wrong ending	8.0	12.2	12.8
Wrong word with correct ending	5.6	5.8	6.4
Other word substitutions	51.2	49.6	46.5
Nonsense (nonword)	6.3	8.4	6.7

Table 3. Graphemic sensitivity of oral reading errors (%)

| Relation of Error | Skill Level | | |
Word to Target Word	Low	Medium	High
No Common Phonemes	32.8	28.1	20.1
Less than 50% Common Phonemes	52.6	48.0	48.7
50% or more Common Phonemes	14.7	23.9	31.3

A general index of graphemic sensitivity can be constructed by classifying errors in terms of the overlap of sounds (phonemes) between the incorrect word uttered and the target word that should have been uttered.* Table 3 shows the distribution of this index, summing over the first five test points. Clearly, the low-skill children's data show markedly less graphemic sensitivity, as indicated by their high rate of producing words with no overlapping letters. However, one-fifth of the error words in the high-skill group were non-overlapped as well, perhaps because of instructional emphasis on attention to context rather than only to graphemic structure of the words.

Examination of the extent to which error words were sensible in context further suggests such an orientation in the instruction. As shown in Table 4, the vast majority of error words in all three skill groups provided a completely sensible reading of the text. The high-ability children more often produced words that made no sense in context than the other children. This is in part because of automatic, but incorrect, recognitions that were later corrected but still tallied as errors. More significant is the greater proportion of partially appropriate items in the low-skill group, a sign that when context did play a role for them, it was a less precise one than for better readers.

---

*For example, cat and mat overlap in two out of three sounds.

Table 4.   Contextual appropriateness of oral reading errors (%)

Relation of error word to target word context in passage	Skill Level		
	Low	Medium	High
Inappropriate	8.8	9.1	16.9
Partly appropriate	20.5	14.0	12.8
Completely appropriate	70.7	76.9	70.3

Searching for the Causes of Reading Disability

The general pattern of the results discussed above suggests that individual word processing skill is an important component of high levels of reading achievement in the primary grades. This is in accord with previous findings on the relations between word processing automaticity and reading skill reported in the literature. Our interest, however, is in going beyond correlation to cause. We know from prior research, as well as from the data in the present study, that children with reading comprehension difficulties lack individual word skills, especially automaticity. But given only the correlations of any two measures taken at about the same time, it is not possible to draw causal conclusions. In the present case we could not decide whether the word recognition facility is necessary for comprehension to succeed or whether it is the result of exercising a higher level of comprehension skill (i.e., perhaps good readers read more and thus become faster word processors.)

Longitudinal data with multiple measurements on the same subjects at separate times can, however, permit causal hypotheses to be tested. In using the data this way, one looks for associations between a variable measured early in the time sequence and one measured later. Since early-occurring events can cause later ones, but later events cannot cause earlier ones, the causal relations between word skill and comprehension can be inferred by comparing the extent to which early word skill automaticity predicts later comprehension with the ability of early comprehension to predict later automaticity. If one of the associations is reliably greater

than the other, a primary direction of causality can be inferred.*
This is the basic logic underlying a number of longitudinal analysis
methods, including cross-lag panel analysis (Campbell, 1963; Kenny,
1975) and structural equation modeling (Joreskog & Sorbom, 1978).

Structural equation modeling (a form of path analysis) was the
technique chosen in this study because it allows simultaneous assessment of the causal relationships among major variables ("latent
variables") and the contributions of several independent measures
to each of those variables. This allows construction of a coherent
model of overall development rather than a collection of interpretations of individual measures and their relationships.

The process begins with the specification of a set of latent
variables (theoretical process components) and the hypothesized
causal relationships among them. In the present case, causal relationships were hypothesized between word skill automaticity and
subsequent reading comprehension ability. According to our theory,
there should be no causal relationships (or at least much smaller
ones) in the opposite direction--i.e., between early comprehension
and subsequent word automaticity. More specifically, for each test
level we hypothesized a latent variable of speed, to be composed
from the oral reading speed and the word vocalization speed measures
taken at that test point. We predicted that this composite speed
variable would strongly contribute to reading comprehension the
next time the standardized reading achievement test was given. By
contrast, we predicted no significant influence of reading comprehension on subsequent speed measures.

The structural equation modeling program, LISREL (Joreskog &
Sorbom, 1978), uses the correlation matrix of the individual observed measures as the basis for choosing a set of path coefficients
between the hypothesized latent variables and a set of weightings
showing how those latent variables are derived from the observed
measures. The coefficients and weightings chosen are those that
optimize the fit of the model to the matrix of observed correlations

Figure 8 shows the results (the causal path coefficients) of
the structural analysis. The vertical dimension of the figure
represents time (from top to bottom). Speed measures are on the
left and reading comprehension on the right. The actual test dates

---

*If relationships in both directions are significant (or not signi-
ficantly different), however, causal inferences cannot easily be
made. Further, while longitudinal data can test specific causal
models, only an intervention study in which the presumed causal
skill is trained and the predicted effect produced can firmly validate the causal prediction.

for the speed measures varied with rate of progress (see Figure 4).
As a result a decision had to be made concerning the place in the
time line for each composite speed variable. Our choices reflect
the underlying distributions of level completion dates and a con-
servative approach of locating levels in ways that did not improve
the chances of confirming our hypotheses.

There are strong causal paths (indicated by solid arrows) between each speed variable and a subsequent comprehension variable,
but no paths of any weight from comprehension to subsequent automaticity. The negative path (indicated by a dotted line) from
Speed at Level 1 to second grade reading comprehension appears to
be due to the speed measure having multiple underlying components.
It has the effect of correcting for influences of word speed at
Level 2 that are due to competences acquired very early in the
curriculum, or before entering school. If the path is removed
from the model, the path from Level 2 Speed simply shrinks a little
bit; it remains substantial. A variety of alternative models were
tried to test the robustness of this pattern. No model that fit
even reasonably well produced path coefficients that were qualitatively different from those of Figure 8; there are no possible
paths left out of the figure that were of any magnitude.

The overall pattern of results provides confirmation of the
assertion that automaticity is a cause of more adequate overall
reading skill and not an epiphenomenon produced by the fact that
better readers practice reading more.

Other features of the automaticity analysis should also be
noted. There are strong paths from each automaticity variable to
the next one in time, indicating a relatively stable automaticity
"trait". The relation between early and later reading comprehension
is less clear. There is a nonsignificant path from the first to
the second grade test, indicating that weak performance on a standardized test of this kind early in first grade does not (in the
absence of word processing slowness) imply weak performance later.
However, second grade reading comprehension seems to contribute as
much or more to third grade comprehension as automaticity does.
One possible reason for this pattern is that first grade comprehension tests are qualitatively different from those that are used
once children can read enough to do serious text processing. In
any case, the implication is clear that early weakness in word
automaticity is a more serious signal of later comprehension difficulties than is poor performance on early standardized tests.

A similar analysis was performed using accuracy instead of
speed measures, in order to determine whether the results in Figure
8 reflected variation in general word recognition skills or more
specific automaticity effects on comprehension. Each ACCURACY

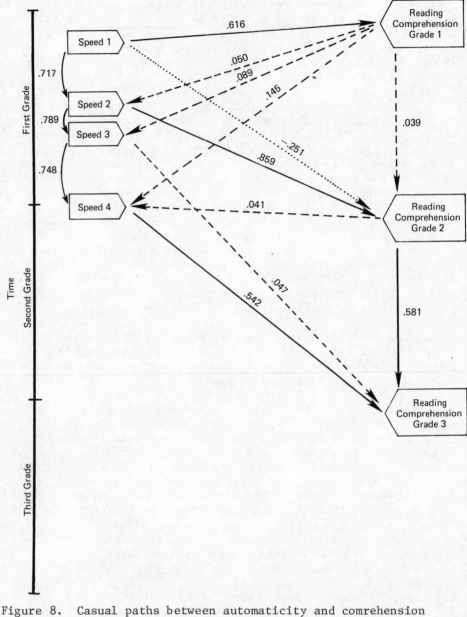

Figure 8.   Casual paths between automaticity and comrehension
            variables.

variable was composed of oral reading and word recognition accuracy measures from a single testing level.   The results of this
accuracy analysis are similar to those for automaticity but show
more independence of word-processing and reading comprehension

skills. The paths from early accuracy to later comprehension are stronger than from comprehension to accuracy, but both the differences between them and their absolute magnitudes are small. Thus, it appears to be efficiency of word processing that has the clearest causal influence on comprehension achievement, at least for the children in our basal curriculum sample.

Implications of the Study

What can be said about the origins of reading disability from the data examined so far? First, it is clear that there exists in the primary classrooms studied a significant minority of children who are reading so poorly by third grade that they will require special attention from the schools if they are to learn to read at functional levels. It is important to recognize, however, that this group of low-skill children is very heterogeneous. The only thing we know that they have in common is weak reading skill. The etiology of their difficulties is not known, nor can we tell which special education label some of these children will acquire as the schools attempt to define appropriate services for them. Furthermore, these problem cases cannot be identified unambiguously on the basis of standard tests given when they began first grade.

Although IQ scores on these children were not available, it appears that some were generally weak in all aspects of school-related skills. Their difficulties do not seem to be specific to reading. Consider the pattern of their scores on the standardized tests taken at the beginning of first grade. As can be seen from Figure 1, it is by no means clear that the low-skill children would generally be classified as disabled readers according to the widely used "discrepancy" or underachievement definition of learning disability. After all, they scored around the fiftieth percentile in reading. Nor do their scores necessarily show markedly uneven profiles across the various subtests.

What, then, has been learned from this body of data that might guide instructional planning in a way that could lower the frequency and severity of reading difficulties that become apparent only in the course of instruction? It seems clear that, while the low-skill children were weak in every component of reading that was tracked, they were especially weak in individual word recognition skills. For example, their graphemic sensitivity was extremely weak while their contextual sensitivity was reasonably good, given the minimally structured texts used. The low-skill children tended to score somewhat lower than others on the Sight Vocabulary, Auditory Discrimination, and Grapheme-Phoneme subtests at the beginning of first grade, but they showed even more marked deficits on the word processing tasks. They were slower and less accurate both in reading ordinary texts and in speeded recognition of individual words, yet they were not slower in general reaction time or in visual letter-search tasks. It was something about recognizing words rather than simple speed of reaction in general that was causing difficulty for these children.

The test battery did not include a strong independent measure of general language skill, especially at the early test points, so poor oral interpretive language skill cannot be eliminated as a cause of reading difficulty. However, it is clear from the pattern of performance on the first grade comprehension subtest that the low-skill group contained some children who apparently had adequate language skills for their age. Thus, while general language deficiencies may play a role in the reading difficulties of some children, it cannot be concluded that the absence of such deficiencies insures smooth acquisition of reading skill. On the other hand, it can be said that the absence of fast and accurate word recognition skills, developed early in the course of learning to read, will almost surely result in deficient reading comprehension ability later. This is the clear import of the path analysis results.

A SAMPLE OF CHILDREN IN A CODE-EMPHASIS PROGRAM

We have also included a sample of children in a direct-instruction, code-emphasis program in our longitudinal study. This offers the opportunity to explore whether the word skills of weak or potentially weak readers show a different pattern of development in a code-emphasis curriculum. The data on the code-emphasis cohorts are still being collected. However, one group of 40 children in the code program was part of one of the earliest cohorts and a complete set of data through third grade is available for them. Although this subgroup is too small for causal modeling, comparisons can be made among low-, medium-, and high-skill groups on the various measures. A fuller report of these comparisons (for first and second grades) has been given by Lesgold and Curtis (1981). We will discuss here only those measures for which a pattern different from the basal instruction cohort can be detected.

The racial and sex breakdowns of the code emphasis were very similar to those for the basal cohort. Their instructional program, the New Reading System (NRS) developed by Beck (Beck & Mitroff, 1972), is organized for individual pacing on the basis of mastery tests that are embedded directly in the curriculum. The core instruction in the first part of Grade 1 is teacher-led, with the children organized into small groups. Thereafter, primary instruction is via tapes and workbooks, so the students are free to progress at very individualized rates. Readers, games, and a variety of related activities provide extensive practice on a

vocabulary of words that is chosen to highlight the grapheme-
phoneme regularities of English, and to provide systematic practice
on specific spelling patterns as they are introduced. Although
the code is stressed, there is also much emphasis on deriving
meaning from text. The extensive practice on regularly spelled
words is also intended to develop a good sight-recognition vocabulary.

Data Patterns for the Code-Emphasis Cohort

Using the same criteria as for the basal reading cohort, 12
of the children in the sample were identified as low-skill readers,
10 as high-skill, and 18 as medium-skill.

Word Processing Speed and Accuracy. The word processing accuracy levels of the code-emphasis cohort were generally higher than
in the basal cohort. Figures 9 and 10 show speed on the word vocalization and word category matching tasks for the code-emphasis
cohort.* Compared with the basal cohort, these children began as
slower word processors; but all skill groups speeded up over levels
and became about as fast as the basal children. The skill groups
appeared to be converging at later levels, suggesting that more
extensive instruction of the type already provided might have removed the speed deficits of the low-skill group. However, a firm
conclusion on this point must await more complete data.

Oral Reading Speed and Accuracy. Figures 11 and 12 show the
mean oral reading speeds for the three groups over levels in the
curriculum. From these figures alone it is clear that even after
three years oral reading speed was quite low for the low-skill
group--around forty words per minute. As with the basal instruction group, there was also a clear difference in speed for the
three skill groups of the code-emphasis cohort. However, there
was a distinct difference in the overall pattern of reading speed
between the two cohorts. While the basal children seemed to operate at a constant speed across many test levels, the code-taught
children began as slow readers but built up to a reading speed
that was considerably faster than the basal children. As the
materials get harder in successive tests, constant speed of the
basal-cohort pattern can be interpreted as reaching a constant
level of mastery for progressively harder text. In contrast, the
speed increases of the code-emphasis cohort suggest positive transfer effects that make each level's performance better than that
of earlier levels--even for new and harder material.

----

*The extremely low number of children tested in the Low Skill
group at Levels 6 and 7 prompted us to not include their data in
the Figures that follow.

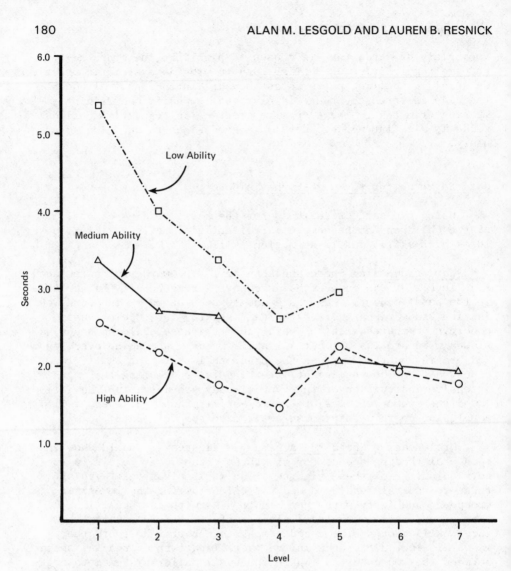

Figure 9. Word vocalization speeds for code-instruction cohort.

The slow reading of the low-skill children in the code-emphasis cohort is all the more marked in the context of the increasing fluency that the medium- and high-skill children showed. The impression one gains is that the instruction in this program encourages the development of word-recognition fluency to a degree that the basal program does not, but that the teachers allowed children to progress through curriculum levels without absolutely insisting on fluency. For this reason some children did not develop this fluency and appeared to be "stuck" as slow readers.

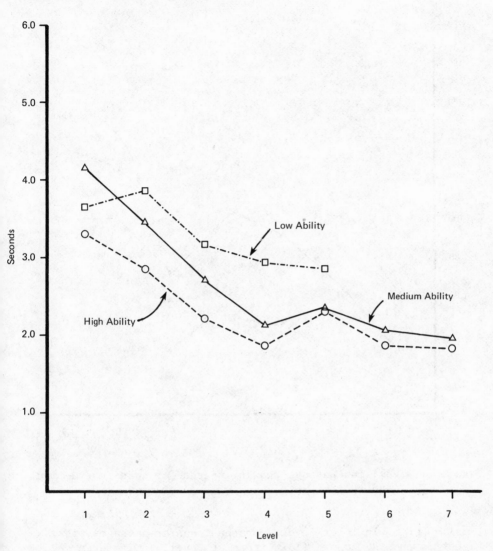

Figure 10.   Category matching speeds for code-instruction cohort.

        This would be possible if the emphasis in teaching were
heavily on decoding accuracy rather than speed. And indeed, this
seems to be the case. Passage from one level to the next in the
NRS program is dependent upon passing a mastery test in which the
child must read a passage aloud while the teacher checks to see
that certain key words are accurately pronounced. There is no
criterion of reading speed, only accuracy. The data on accuracy
reflect this. Oral reading accuracy was generally at least as
high for all groups in the code-emphasis cohort as it was in the

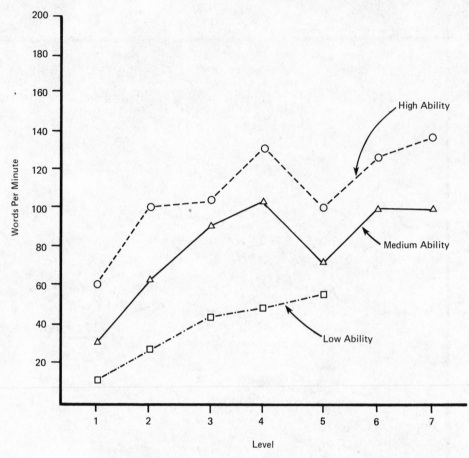

Figure 11. Familiar passage reading speeds for code-instruction
            cohort.

basal cohort. In later levels, on the transfer passages, it was
higher in the NRS cohort (18% errors at Level 4) with even the
low-skill children more accurate than their counterparts in the
basal cohort (23% errors).

    Error Analyses. Error types on the oral reading task were
distributed almost identically to those in the basal cohort (see
Table 2), with the exceptions that the code-emphasis group had
slightly more nonsense errors (12% vs. 7%). The finding of only
a slightly greater tendency to utter a nonsense response does not
seem consistent with anecdotal claims that code-emphasis curricula
produce "word callers" who are completely insensitive to meaning.

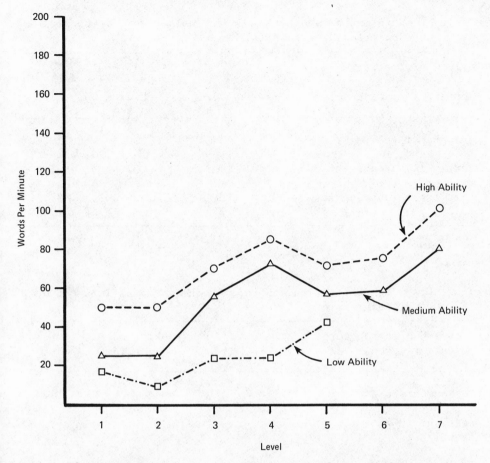

Figure 12.   Transfer passage reading speeds for code-instruction
             cohort.

An examination of contextual and graphemic sensitivity of
errors, however, did show clear curricular differences. Compare
Table 5 for the code-emphasis cohort to Table 4 above. There is
universally high context sensitivity in the basal group. There
is less context sensitivity in the code-emphasis group.* On the
other hand, there is higher graphemic sensitivity in the code-
emphasis group, as shown in Table 6 (compare to Table 3).

---

*Since the child's first utterance was scored, this would tend
to underestimate contextual sensitivity in any case where a
sounding-out strategy was applied i.e., where a word was pronounced and then rejected as not fitting.

Table 5. Word emphasis cohort: Contextual appropriateness of
oral reading errors (%)

Relation of error word to target word context in passage	Skill Level		
	Low	Medium	High
Inappropriate	24.8	20.2	10.2
Partly inappropriate	61.7	57.4	61.6
Completely inappropriate	13.6	22.4	28.2

Table 6. Code-instruction cohort: Graphemic sensitivity of oral
reading errors (%)

Relation of error word to target word	Skill Level		
	Low	Medium	High
No common phonemes	8.5	8.8	12.9
Less than 50% common phonemes	22.6	14.3	13.3
50% or more common phonemes	68.9	76.9	73.9

CONCLUSION

Two quite different instructional programs yielded some different patterns of reading skill acquisition. Neither curriculum,
however, completely solved the problem of the low-skill children.
Children with weak measured comprehension performance remained
very slow readers in both programs. This suggests, in accord with
automaticity theory, that lack of word processing efficiency may
lie at the heart of reading disability.

The obvious suggestion for instruction that emerges from
these findings is that more emphatic and systematic attention to

word recognition skills might reduce or eliminate the number of
children with severe reading difficulties. This proposal finds
some support in comparative studies of code versus sight-word
recognition instruction over a number of decades. Resnick (1979)
has summarized these findings as follows: direct instruction in
a program that emphasizes the grapheme-phoneme code generally improves the reading performance of less able learners; no difference
among programs is generally found for the more able learners. In
accord with these findings, there has been a growing use of code
programs (such as Open Court and Distar) with special education
populations in recent years (Bateman, 1979). The evidence from
this study suggests that an emphasis on efficiency as well as
accuracy of decoding should be added to these programs in order
to build the automated word recognition skill that seems to be
required for normal comprehension skill development. Such an
emphasis, may, of course, require substantially new teaching approaches. Implementation of such modified programs will eventually
provide the data to confirm or disconfirm our hypothesis that
early attention to word-level efficiency will facilitate comprehension.

There is one caveat: we do not know for certain that all
cognitive efficiency differences are remediable by training. At
least in certain more severe cases, a specific speed-of-processing
deficit may be irremediable. But poor reading skill is too common
for us to believe that it is entirely due to factors that cannot
be changed by instruction. If additional instruction is the answer, then direct instruction is most likely to be successful.
The evidence so far suggests that word processing efficiency
should become a locus of such direct instruction.

References

Bateman, B. Teaching reading to learning disabled and other hard-
    to-teach children. In L. B. Resnick & P. A. Weaver (Eds.),
    Theory and Practice of Early Reading (Vol. 1). Hillsdale, NJ:
    Lawrence Erlbaum Associates, 1979.
Beck, I. L., & Mitroff, D. D. The rationale and design of a pri-
    mary grades reading system for an individualized classroom.
    University of Pittsburgh: Learning Research and Development
    Center Publications Series, 1972.
Campbell, D. T. From description to experimentation: Interpreting
    trends as quasi experiments. In C. W. Harris (Ed.), Problems
    in measuring change. Madison: University of Wisconsin Press,
    1963.
Curtis, M. E. Development of components of reading skill. Journal
    of Educational Psychology, 1980, 72, 656-99.

Fleisher, L. S., & Jenkins, J. P. Effects of contextualized and decontextualized practice conditions on word recognition. Learning Disabilities Quarterly, 1978, 1(3):39-74.

Frederiksen, J. R. A componential theory of reading skills and their interactions (Report No. 4649). Cambridge, Mass: Bolt, Beranek and Newman, April 1981.(a)

Frederiksen, J. R. Sources of process interactions in reading. In A. M. Lesgold & C. A. Perfetti, (Eds.), Interactive Processes in Reading. Hillsdale, NJ: Lawrence Erlbaum Associates, 1981.(b

Frederiksen, J. R. Assessment of perceptual, decoding, and lexical skills and their relation to reading proficiency. In A. M. Lesgold, J. W. Pellegrino, S. D. Fokkema, & R. Glaser (Eds.), Cognitive Psychology and Instruction. New York, NY: Plenum Press, 1978.

Goodman, K. S., & Goodman, Y. M. Learning to read is natural. In L. B. Resnick & P. A. Weaver (Eds.), Theory and Practice of Early Reading (Vol. 1). Hillsdale, NJ: Lawrence Erlbaum Associates, 1979.

Hood, J. Qualitative analysis of oral reading errors: the inter-judge reliability of scores. Reading Research Quarterly, 1975-6, 6(4), 577-99.

Jackson, M. D. & McClelland, J. L. Processing determinants of reading speed. Journal of Experimental Psychology: General, 1979, 108(2), 151-181.

Joreskog, K. G., & Sorbom, D. LISREL IV: A general computer program for estimation of linear structural equations by maximum likelihood methods. Chicago: International Educational Services, 1978.

Just, M. A., & Carpenter, P. A. (Eds.), Cognitive Processes in Comprehension. Hillsdale, NJ: Lawrence Erlbaum Associates, 1977.

Kenny, D. A. Cross-lag panel correlation: A test for spuriousness. Psychological Bulletin, 1975, 82, 887-903.

LaBerge, D., & Samuels, S. J. (Eds.), Basic Processes in Reading: Perception and Comprehension. Hillsdale, NJ: Lawrence Erlbaum Associates, 1977.

LaBerge, D., & Samuels, S. J. Toward a theory of automatic information processing in reading. Cognitive Psychology, 1974, 6, 293-323.

Lesgold, A. M., & Curtis, M. E. Learning to read words efficiently. In A. M. Lesgold & C. A. Perfetti (Eds.), Interactice Processes in Reading. Hillsdale, NJ: Lawrence Erlbaum Associates, 1981.

Lesgold, A. M., & Perfetti, C. A. (Eds.) Interactive Processes in Reading. Hillsdale, NJ: Lawrence Erlbaum Associates, 1981.

Perfetti, C. A., & Hogaboam, T. W. The relationship between single word decoding and reading comprehension skill. Journal of Educational Psychology, 1975, 67(4), 461-469.

Perfetti, C. A., & Lesgold, A. M.   Coding and comprehension in
   skilled reading and implications for reading instruction.   In L.
   B. Resnick & P. A. Weaver (Eds.), Theory and Practice of Early
   Reading (Vol. 1).   Hillsdale, NJ: Lawrence Erlbaum Associates,
   1979.
Perfetti, C. A., & Roth, S.   Some of the interactive processes in
   reading and their role in reading skill.   In A. M. Lesgold &
   C. A. Perfetti (Eds.), Interactive Processes in Reading.   Hills-
   dale, NJ: Lawrence Erlbaum Associates, 1981.
Resnick, L. B.   Theory and practice in beginning reading instruction.
   In L. B. Resnick & P. A. Weaver (Eds.), Theory and Practice of
   Early Reading (Vol. 3).   Hillsdale, NJ: Lawrence Erlbaum Asso-
   ciates, 1979.
Resnick, L. B., & Weaver, P. A. (Eds.), Theory and Practice of
   Early Reading (Vols. 1-3).   Hillsdale, NJ: Lawrence Erlbaum
   Associates, 1979.
Rumelhart, D. E., & McClelland, J. L.   Interactive processing
   through spreading activation.   In A. M. Lesgold & C. A. Perfetti
   (Eds.), Interactive Processes in Reading.   Hillsdale, NJ:
   Lawrence Erlbaum Associates, 1981.
Smith, F.   Conflicting approaches to reading research and instruc-
   tion.   In L. B. Resnick & P. A. Weaver (Eds.), Theory and
   Practice of Early Reading (Vol. 2).   Hillsdale, NJ: Lawrence
   Erlbaum Associates, 1979.

READING AND SPELLING DIFFICULTIES

Lynette Bradley and Peter Bryant

Department of Experimental Psychology
Oxford University, Oxford, England

One of the most reasonable and least controversial things to be said about reading and writing is that these are very complex skills indeed. They clearly involve a large number of separate psychological processes. If he is to read, a child must be able to dismember words phonologically, to recognize and tell apart visual configurations, to learn rules about visual to auditory transformations, to cope with the many exceptions to these rules, to remember sequences and so on.

Apply these complexities to the question of children who do not manage to learn to read or write properly, and one is immediately faced with two problems. First, there may very well be differences between these children: different backward readers might experience different types of difficulty, some with phonological codes, for example, and others with visual memory. Secondly, simply to give a shopping list of the processes involved in reading is not enough on its own. We really ought to be able to say something of how these processes work - how, that is, phonological codes and visual codes interact in reading and spelling. We ought to know about this sort of thing so that we can do more than just say of any individual child with reading problems that he has this or that weakness. We ought also to be able to decide exactly what effect the weakness has on his attempts to learn to read.

Put like this, it is easy to see that the research that exists on reading difficulties reflects our lack of knowledge about what exactly is involved in reading and writing script. This research has concentrated on detecting the backward reader's weaknesses. There is now an immense shopping list of these. However, the precise effects of the weaknesses are not at all clear. What can be

done about this? The answer seems to us to start with the weakness
and then to track down its effects on the actual business of reading
with a combination of intensive studies of individual cases and
experiments both with normal and with backward readers.

Let us take as an example the question of phonological codes.
Nothing could be more obvious than the need for some kind of phonological awareness in learning to read and write. The alphabet works
by breaking words down into their constituent sounds, and it seems
plausible to suggest that any child who cannot break words up in
this way is going to have some difficulty mastering the alphabetic
code and, therefore, the rudiments of written language.

This is not a new point and there are several people who have
tried to demonstrate a phonological weakness in backward readers.
That their attempts are entirely justified can be neatly illustrated
by the study of individual cases. We should like to describe one
particular boy who presented an extreme example of this sort of difficulty. He was 16 years and one month old when we saw him. His
intelligence level and his vocabulary were normal and yet he had
made virtually no progress at all with reading, and even less with
spelling. (His reading and spelling ages were 7 years-4 months and
6 years, respectively.) He turned out to have difficulty with any
task that involved manipulating sounds, and he was particularly
confused at the level of the phoneme. At first, he also had some
difficulty with dividing words into syllables. When we gave him
words like 'remember' or 'carpet' and asked him to tell us how many
syllables they contained he could not, to begin with, give us the
correct answer. However, he quickly put that right within one
session.

His problems in analyzing words into phonemes were more
obdurate. One clear sign of this was that this normally intelligent
boy of 16 did not have the correct label for letters and in particular confused letters which represented similar sounds, e.g. g and k.
He had, as might be expected, tremendous difficulty deciphering
written nonsense words, particularly when these involved more than
one syllable. Another sign was the way he attempted to spell words.
Often he would start by writing the last sound (starting 'then' with
'n', for example). The most striking illustration of his difficulty
with phonological analysis came in tasks that had nothing directly
to do with reading at all. We wanted to know how good he was at
spotting whether different words had sounds in common, and we devised various tasks to see whether he could tell us when words ended
or began with the same sounds. We also looked at his ability to
produce such rhymes or alliteration. On both sides - input and output - he was very weak indeed, and on some tests did a great deal
worse than what we know to be the level for the average child of
five years. He had a phonological problem. This in fact was not
his only difficulty since we demonstrated later that he also had a

poor visual memory. But clearly his main weakness was with sounds.
He could not analyze them or blend them.

　　　How general is this sort of problem among backward readers? As
we have said, many psychologists have suggested that it is frequent.
However, the empirical evidence is often not all that impressive.
For one thing, it usually comes from experiments which make the
wrong comparison. They pit groups of backward readers against
normal readers with the same chronological and mental ages, the only
difference between the two groups being in their reading levels.
The difficulty with this design is that any difference that emerges
between the two groups could as well be the result of the reading
problems as their cause. There is a simple solution which is to
match backward and normal readers on reading age rather than on
chronological age and mental age. We wish to report a large scale
study (Bradley & Bryant, 1978) in which we included a control group
matched on reading age and yet showed a striking difference between
the backward and normal readers. The study was divided into two
parts. In the first (the input part), the words were read out to
the children four at a time. Three of the words had a phoneme in
common which the fourth did not share. The children's task was to
spot the odd one out. Sometimes the sound in question was the beginning phoneme (sun, sock, see, rag), sometimes the middle vowel
(nod, red, fed, bed) and sometimes the final consonant (weed, peel,
need, deed).

　　　All three input tasks, as Table 1 shows, were immeasurably more
difficult for the backward readers. They could remember them
alright, but they could not, apparently, analyze them into phonemes.

Table 1. Mean error scores (out of 6) when detecting rhyme and
　　　　　　alliteration

Series	Odd word	Backward readers N = 60		Normal readers N = 30	
		Mean	s.d.	Mean	s.d.
1	Last letter different	1.15	1.43	0.17	1.11
2	Middle letter different	1.49	1.58	0.37	0.99
3	First letter different	2.62	2.26	0.67	1.188

Table 2.   Number in each group failing to produce rhymes

	Total N	No. of failures										
		0	1	2	3	4	5	6	7	8	9	10
Backward readers	60	37	4	4	4	2	2	2	2	0	0	3
Normal readers	30	28	1	0	0	0	1	0	0	0	0	0

There was further evidence of this weakness in the second part of the study - the output part. The children were asked to produce a meaningful rhyme for ten words. Nearly all the normal readers could manage this task perfectly: not so the backward readers who made many mistakes, as Table 2 shows.

To find such clear differences twice over - and this despite the fact that our design demanded that the backward readers had a considerably higher mental age than the normal controls - does seem to indicate a frequent phonological difficulty among backward readers. However, is this all one needs to say? Our equation so far goes as follows. Reading and spelling depend on the alphabet: the alphabet works by breaking words up phonologically: backward readers are weak at precisely this kind of phonological segmentation: therefore, they become backward readers. It is neat enough, but almost certainly too simple.

Our problem is that many words which even very young children learn to read are not easy to construct phonologically letter by letter. 'People' and 'school' are just two examples. It is possible that children learn to read them in some other way, by learning sequences of letters or recognizing whole word shapes, for example. Thus, the phoneme by phoneme analysis, which we seem to be showing to be so difficult for the backward readers, may not be necessary for reading many common words. So we need to ask, as we warned at the beginning, not just what the backward reader's weaknesses are, but precisely what effect these weaknesses have on learning to read and to spell.

Let us start an answer with the obvious, though often neglected, point that reading and spelling are very different activities and may be done in very different ways, so that anything we find about reading might not apply to the way children spell. What can be said about the role of phonological segmentation in both activi-

ties? Strangely enough, there is very little direct evidence on
this question, at least as far as young children are concerned.
One of the most impressive pieces of work actually suggests very
little involvement of phonological segmentation in reading single
words. This was a study by Barron and Baron (1977) who used a technique known sometimes as articulatory interference, sometimes as
phonological interference and sometimes more prosaically as concurrent vocalization. The aim of this technique is to suppress all
phonological activity and to see if reading is still possible.
Their version of the technique was simply to give children a list of
picture word pairs and to ask them to judge whether the word meant
the same as the picture (sometimes it did and sometimes it did not):
in one condition - the interference condition - the children had to
repeat the word 'double' 'double' 'double' all the while; in the
control condition, the child was silent. The question they asked
was whether the interference would harm the children's reading, and
the evidence was that it did not at all. Even beginning readers
read as fast and with a few mistakes when they had to keep on saying
'double' as when they did not. The authors concluded from this surprising result that the children could not have been reading by constructing the word phonologically letter by letter and sound by
sound.

    If this conclusion is right, we have a serious problem. It
would mean that backward readers have difficulty with phonological
segmentation and yet, at the same time, phonological segmentation is
not necessary for reading. Why, then, are they backward? There is,
however, a way out which is to argue, as some have, that the so-
called phonological interference does not in fact stop phonological
activity. Phonological activity might still persist in the form of
central nervous activity even when the articulatory system itself
is otherwise engaged. One way to test this hypothesis is to give
children reading tasks which involve juggling with phonological segments. If Barron and Baron's (1977) argument is right, these would
be affected by interference. On the other hand, according to the
counter-argument that 'double' 'double' doubling has little effect
on the phonological system, juggling with phonological segments
should be as easy under interference as reading whole words.

    So we conducted as experiment with six- and seven-year-old
normal readers in which there were four kinds of tasks. One of
these, the whole word task, was similar to Barron and Baron's (1977)
task. The children were given ten words, each paired with a
picture; in five pairs the word meant the same thing as the picture
while in the other five it did not. The child's task was to mark
those pairs in which the meanings of the pictures and words were
the same. In fact, two variations of the whole word task were
given, one involving easy words and the other involving difficult
words.

The next three tasks were all designed to test the child's
ability to deal with segments and, therefore, involved single
letters rather than whole words. The first of these was the first
letter task. Again the children were given ten pictures and again
beside each picture was a three letter word; but now, the second and
third letters were completely and very explicitly blacked out. The
child's task was to judge whether the remaining letter, the first
one, represented the picture's first sound. The second letter task
and the third letter task were similar except that the single letter
remaining after the other two had been blacked out was either the
middle letter or the last letter.

We gave six- and seven-year-old children, all normal readers,
these four tasks under two conditions. One involved interference or
concurrent vocalization - the children had to say 'bla-bla' all the
while. The other was a silent condition. The results were unambig-
uous. The interference really did interfere with the single letter
task; it had absolutely no effect at all on the whole word condition, as Table 3 shows. So these results support Barron and Baron's
(1977) original and surprising conclusion very strongly. The difference between the whole word and the single letter tasks does
suggest that the children were not reading the whole words in a
letter by letter phonological manner.

If one follows this conclusion through, one might be tempted
to think that the phonological impairment that featured so
strongly in our rhyming studies actually had nothing to do with the
backward reader's difficulties with written language. But from a

Table 3. Mean correct scores (out of 10) in the whole word and
          single letter tasks

		Whole Word		Single Letters		
		Easy	Difficult	First letter	Middle letter	End letter
6 year-olds	Silent	9.2	6.5	8.5	6.9	7.1
	Concurrent Vocalization	8.9	6.8	6.4	4.9	4.9
7 year-olds	Silent	9.6	7.5	9.9	7.5	7.8
	Concurrent Vocalization	9.6	7.9	7.2	5.1	5.8

Table 4.   Mean correct spelling scores (out of 10)

		Easy	Difficult
6 year-olds	Silent	4.9	3.4
	Concurrent Vocalization	2.0	0.7
7 year-olds	Silent	5.2	3.1
	Concurrent Vocalization	3.9	1.4

different point of view, one may argue that the other main activity in written language is writing. It is possible that children read in one way and spell in another. Therefore, it is possible that even if phonological segmentation is not an important part of reading single words, it nevertheless plays an important role in children's spelling. This is, in fact, exactly the hypothesis that we shall now put forward.

We have a simple and effective way of testing this hypothesis: give the same children a spelling task with and without phonological interference. If our hypothesis is correct, the interference which leaves their reading unscathed should have a devastating effect on their spelling. This was exactly what happened. We gave some six- and seven-year-old children sets of six pictures and they simply had to write the name of each picture down beside it. There were easy and difficult to spell lists (the lists given to the seven-year-olds were harder, in absolute terms, than those given to the six-year-olds). They did one of each of these lists under phonological interference (again saying 'bla bla' all the while), and the other silently. Both the easy and difficult lists were a great deal harder under interference, as Table 4 shows. Indeed there were children who spelled quite well in the silent condition but who simply could not cope when interference was introduced.

We concluded that, for the beginning reader at any rate, phonological analysis is much more important in spelling than in reading. Thus, we have at least found one activity to do with written language which actually does demand some phonological segmentation, and so we can see why backward readers' phonological weaknesses might have some effect.

So we are arguing for some separation between reading and
spelling in the beginning reader, and also in the backward reader.
As yet, we have not reported a direct comparison in a single experiment. However, we do have some direct evidence which provides convincing support for our notions about the early independence between
reading and spelling (Bradley & Bryant, 1979; Bryant & Bradley,
1980). Children - the same backward and normal readers who were in
the rhyming study described earlier - were given a list of words to
read and to spell on separate occasions. The question then is: Do
the children read and write the same words? Are the words which
they read also the words which they tend to spell correctly? The
words can be divided into four categories for each child: RS words
which he both reads and spells correctly. R̄S words which he gets
wrong both in reading and in spelling, RS̄ words which he manages to
read but does not spell properly and finally the fourth possibility,
R̄S words which he does not read but manages to spell correctly.

The important categories here are the last two, which we can
call the discrepant categories. If we can find a reasonable number
of examples of both kinds of discrepancy between reading and
spelling we would have evidence of a separation between reading and
spelling. If there are words which a child reads but does not spell
and others which he spells but does not read, it is likely that, at
least for some of the time, he is reading and spelling in different
ways. The greater the number of words that fall into these discrepant categories, the more striking is the independence.

The results of this study are very simple. Both groups showed
both kinds of discrepancy, as Table 5 shows, but the proportion of
words which fell into the discrepant categories was greater in the
backward readers, even though, it must be remembered, the reading
levels of the two groups were the same. So we do, by means of the
simple but surprising observation that children often spell words
which they do not read as well as read words which they do not
spell, have good evidence of a separation between reading and spelling in young children and, to a larger extent, in backward readers.

This takes us to the next stage of our analysis which is to
see whether the different ways in which they set about these two
activities coincide with our idea of spelling being primarily phonological and reading primarily visual.

As a start we looked at the form of reading and spelling
errors, to see what phonological connections there were between the
mistakes and the correct word. Our analyses showed that the mis-
spelled words had more sounds in common with what would have been
the correct spelling than the mis-read words had in common with what
would have been the correct reading, as Table 6 shows. So we do
have initial support for the idea of the phonological strategy for
spelling.

Table 5. Mean number of words (out of 18) in the four possible categories

	N	RS Words read and spelled	$\overline{R}\,\overline{S}$ Words neither read nor spelled	$R\overline{S}$ Words read but not spelled	$\overline{R}S$ Words not read but spelled
Backward readers	62	7.7	4.9	3.1	2.3
Normal readers	30	10.6	3.9	2.1	1.4

We also have similar evidence from another experiment. Two age groups of children (all normal readers) were again given words to read and to spell on different occasions. The younger group produced an appreciable number of both kinds of discrepant words, thus again demonstrating the independence which interests us. Interestingly enough, as Table 7 shows, one of the two discrepant categories, the surprising category of words which are spelled but not read, virtually vanished in the older group - a developmental change which makes us think that the separation between reading and spelling which we have found is a transitory beginning phase for most children.

The main point of this study lies in the further analysis of the younger group's results. First of all, looking again at the two discrepant categories, it is quite clear that very different words fall into them. The words spelled but not read tend to be easily constructed phonologically: the most common instances were 'bun,' 'mat,' 'leg,' 'pat.' Not so the words read but not spelled. The commonest were: 'school,' 'light,' 'train,' 'egg.' It is easy to see that these last four words are extremely difficult to construct on a simple letter by letter basis.

Taken together, the results of these two studies enable us to state that the initial independence of reading and spelling is normally a transitory phenomenon. The two main strategies - visual and phonological - come together in both activities. However, in backward readers they stay separate, even when reading levels are taken into account.

Table 6. The connection between the correct answers and the mis-spelled and the mis-read words

	No connection	Same first sound	Same last sound	Same first and last sound	Same first, and intervening sound	Same first, and intervening and last sound	Same intervening and last sound
Normal readers							
Mis-spelled words (%)	4.6	4.6	0.8	6.6	31.6	44.8	7.9
Mis-read words (%)	10.0	13.9	0.8	12.3	26.9	23.1	13.1
Backward readers							
Mis-spelled words (%)	3.4	2.9	0.7	12.7	23.6	46.6	10.1
Mis-read words (%)	13.6	11.5	0.6	10.6	29.7	25.7	8.2

Table 7.   Mean number of words (out of 30) in the four possible
           categories

	N	RS Words read and spelled	$\overline{R}S$ Words neither read nor spelled	$R\overline{S}$ Words read but not spelled	$\overline{R}\overline{S}$ Words not read but spelled
Younger group	30	12.7	7.1	6.3	3.9
Older group	20	18.5	5.5	5.5	0.5

Thus, we see two kinds of problems within backwardness in
reading, and we think a sharp distinction between the two is needed.
The first problem involves a weakness in a necessary strategy in
reading and spelling. We have stressed the case of weaknesses in
phonological strategies, although we are aware of the possibility of
other weaknesses as well. The second is a problem of a lack of coordination of different strategies. This kind of problem has not
been stressed before, and we feel it deserves attention.

Conclusion

For our conclusion we return to our case history. Here was a
boy who experienced difficulty with one essential strategy, a
strategy which, we have shown, children seem to use straightaway
when they spell, though only later in their reading. Not surprisingly, his spelling had not even begun, whereas his reading had
made a little progress and then stopped.

We feel that cases like his would be helped by a proper recognition both of the difficulties caused by a lack of phonological
awareness, and also of the fact that reading and spelling involve a
number of different strategies. Both things are relevant to the
treatment of backwardness in reading, as indeed we have found in
this case. Our recommendation is a two part one. Bolster the weak
strategy, and at the same time try to get the child to take as much
advantage as he can of those strategies of his which are unimpaired.

References

Barron, R.W., & Baron, J.   How children get meaning from printed
   words. Child Development, 1977, 48, 587-594.

Bradley, L., & Bryant, P. E.   Difficulties in auditory organization as a possible cause of reading backwardness. Nature, 1978, 271, 746-747.

Bradley, L., & Bryant, P. E.   The independence of reading and spelling in backward and normal readers. Developmental Medicine and Child Neurology, 1979, 21, 504-514.

Bryant, P. E., & Bradley, L.   Why children sometimes write words which they cannot read. In U. Frith (Ed.), Cognitive processes in spelling. London: Academic Press, 1980.

COMPUTER-AIDED LEARNING: PERFORMANCE OF HYPERACTIVE AND LEARNING

DISABLED CHILDREN[1]

Robert M. Knights

Department of Psychology
Carleton University, Ottawa, Ontario, Canada

This chapter provides a brief review of the literature on computer-aided learning (CAL) with exceptional children. The main focus is directed toward the influence of the type of feedback and the nature of the task involved. A study is presented which investigates differential feedback effects in hyperactive and learning disabled children.

Following the tremendous growth of computer use in the business community, professionals in the areas of psychology and education are gradually realizing the usefulness of the computer for teaching and training. In the last decade many CAL laboratories have been developed. Stanford University has been involved in teaching reading (Atkinson, 1968, 1974; Fletcher & Atkinson, 1972) and mathematics (Suppes & Searle, 1971) to elementary school children and the University of Illinois' PLATO system provides a wide variety of programs. In addition, many other centres in both Canada and the United States are becoming well known for their programs in CAL.

There is evidence to suggest that with certain subgroups of children, automated assessment and automated training programs may be more effective than regular procedures. For example, black children score higher on automated ability tests (Johnson & Mihal,

[1]This research was supported by the Social Sciences and Humanities Research Council of Canada, Grant Number 410-77-0695. The author appreciates the assistance of June Cunningham, Kathy Sheehan, Clare Stoddart, Pat Thompson and Jan Norwood in the preparation of this paper.

1973) as do culturally deprived children (Feldman & Sears, 1970;
Fletcher & Atkinson, 1972). Preliminary studies have suggested that
the computer is most useful when used as a tool to present practice
and drill exercises, rather than for actual teaching or presentation
of new material (Rockart, 1974; Seltzer, 1971). It has also been
suggested that CAL programs may be particularly beneficial to the
slow learning child as they allow the child to proceed at his own
pace. This suggestion has been supported by Freibergs and Douglas
(1969) who found that hyperactive children may learn as well as
normal children, if given the opportunity to proceed at their own
pace. Fahlam (1965) and Hartley, Lovell, and Sleeman (1972) found
that students who began CAL programs with low ability gained more
from computer instruction than did students with greater ability.
This paper is concerned with particular subgroups of children who
may also benefit from the advantages of CAL.

FEEDBACK AND CAL

     One component considered to be of prime importance in pro-
gramed learning is that of feedback, or knowledge of results. Most
CAL programs designed to record and evaluate subject responses
incorporate some form of feedback. In spite of this, the effects of
feedback are rarely studied. The research that has dealt with this
question is most often concerned with the type of feedback. Gilman
(1967, 1969) investigated this variable in a CAL program for
teaching science concepts to secondary school and university students. With high school students, he found no difference in learning when the correct answer appeared alone after each response, or
when the correct answer was presented along with response-contingent
feedback (right or wrong) after each response (Gilman, 1967). In
contrast, for university students (Gilman, 1969) the increased
information given by presenting the correct answer with or without
direct feedback facilitated the learning task. It is suggested that
Gilman's (1969) study is better designed and therefore more valid
than the 1967 study. However, this issue has not been resolved.

     Tait, Hartley, and Anderson (1973) conducted a similar study of
feedback. They found that grade 2 and 3 children who were presented
feedback after responses, as well as subsequent problems according
to the correctness of their responses, learned simple multiplication
faster than children who progressed through the program according to
the correctness of their responses but without feedback. Similarly,
Majer, Hansen, and Dick (1971) found that high school students
performed better when some form of feedback was presented (either
positive verbal statements or simple indications of correct responding) than when they simply progressed through the program.

THE EFFECTS OF FEEDBACK ON NORMAL CHILDREN'S LEARNING

The effects of feedback on normal children's learning have been studied extensively, most often with the use of a simple two-choice discrimination learning task. The typical experimental paradigm involves the presentation of trials under three feedback contingencies (for example, Spence & Segner, 1967). The first condition involves presentation of a positive feedback event or positive reinforcement, for example, the word "Right" after correct responses and no feedback after incorrect responses (the R∅ or right-blank condition). In the second condition, negative feedback or negative reinforcement, for example, the word "Wrong" is presented after incorrect responses and no feedback occurs after correct responses (the W∅ or wrong-blank condition). Finally, in the third condition, positive feedback for correct responses and negative feedback for incorrect responses is administered (the RW or right-wrong condition). These studies have been conducted with tangible or material reinforcements such as tokens, candies or tone presentations (Brackbill & O'Hara, 1958; Penney, 1967; Penney & Lupton, 1961; Stevenson, Weir, & Zigler, 1959; Tindall & Ratliff, 1974; Whitehurst, 1969; Witte & Grossman, 1971) and with verbal feedback, such as the words "right" or "good" and "wrong" or "no" (Curry, 1960; Meyer & Offenbach, 1962; Meyer & Seidman, 1961; Ochnocki, Cotter, & Miller, 1974). Results with both verbal and material feedback indicate that the use of negative feedback contingent on incorrect responses is more effective than the use of positive feedback contingent on correct responses, that is, the W∅ and RW conditions are more effective than the R∅ condition for children's learning. Results are inconclusive as to whether negative feedback contingencies are more effective when used alone (W∅) or when used in combination with positive feedback (RW).

Many of these investigators have presented a theoretical framework with which to explain the differential effectiveness of the R∅, W∅ and RW contingencies. Basically, there are two models. The motivational model suggests that the event ∅ is a non-reinforcer, whereas the events R and W act as positive and negative reinforcements, respectively. The fact that the W∅ condition is more effective than the R∅ condition is explained by assuming that R is weaker on the reinforcement continuum, or produces less incentive value than W (Buchwald, 1959; Buss, Braden, Orgel, & Buss, 1956; Schmeck & Schmeck, 1972). The feedback, whether R or W, acts to motivate the subject towards or away from a particular response.

In contrast to this model, the informational model first advocated by Spence (1964) suggests that the R and W events facilitate learning by providing information, that is, strictly feedback with no mention of reinforcing or motivational properties. She attributes the difference in learning under R∅ and W∅ to the notion

that the ₿ event acquires informational properties during the learning session, but does so more slowly with less informational value
when combined with "Right" (R₿) than when combined with "Wrong"
(W₿).   The rationale underlying this assumption is that children are
more frequently told when they are wrong than when they are right.

Whether an informational or a motivational model best explains
learning behavior has not been decided.   Recent research in the
author's laboratory suggests that different types of subjects may
demonstrate learning according to the different models suggested.
For example, Fike (1977) found that mentally retarded adolescents
performed in a computer-assisted discrimination learning task
according to the informational hypothesis.   In a study involving the
variable of socio-economic status (SES) in computer-administered
discrimination learning, it was found that the performance of
middle-class children was best explained by the informational hypothesis, whereas the low-middle SES children were affected by the
feedback according to the motivational hypothesis (Babcock, 1977).
These models may serve as important indicators for the learning of
hyperactive and learning disordered children and are further investigated in the research presented.

THE EFFECTS OF FEEDBACK ON HYPERACTIVE AND LEARNING DISORDERED CHILDREN

There are reports in the recent literature suggesting that the
hyperactive child, and perhaps the learning disabled child, is different from the normal child in his reaction to the effects of
positive and negative reinforcement and to different schedules of
feedback.

Studies from Douglas (e.g., 1972) and her group of investigators at McGill University have demonstrated that, at least with
some feedback conditions, the hyperactive child takes longer to
learn, or performs with less accuracy than the normal child.
Freibergs and Douglas (1969) and Parry and Douglas (1974) have shown
this to be the case in a concept learning task.   They demonstrated
that the hyperactive children took longer to learn under a partial
reinforcement schedule.   However, there was no difference between
the hyperactives and normals under continuous reinforcement.   These
effects have also been demonstrated in a delayed reaction time task
(Cohen & Douglas, 1971), where it was shown that when extinction was
instituted after a continuous reinforcement schedule, the normal
children were able to maintain the improved level of performance
much longer than the hyperactive children.   In contrast, Firestone
(1974), using a similar delayed reaction time task, found no differential effects of three types of feedback, positive (R₿),
negative (W₿), or both positive and negative (RW), in the performance of hyperactive and normal children.

A number of studies concerned with the effects of feedback with reflective and impulsive children are relevant to hyperactive children. Campbell, Douglas, and Morgenstern (1971) found hyperactive children to be impulsive as defined by Kagan's (1965) Matching Familiar Figures Test of reflection-impulsivity. In a discrimination learning task, Hemry (1973) demonstrated that reflective children performed better than impulsive children, regardless of the reinforcement procedure. There were no effects of type of feedback (verbal or non-verbal) and no differential effects of the feedback condition between the reflective and impulsive children.

In contrast to Hemry's research, both Firestone and Douglas (1974) and Massari and Schack (1972) concluded that negative verbal feedback contingent on incorrect responses not only resulted in better performance for reflective and impulsive children but also caused the impulsive children to respond more like the reflective children. In fact, in the study by Firestone and Douglas (1974), there was no difference between reflective and impulsive children's performance under negative feedback contingencies.

In another study using tangible reinforcements, Cunningham and Knights (1978) administered an automated two-choice discrimination learning task to older and younger groups of hyperactive and normal boys. Both hyperactive and control group children learned the task faster and showed greater resistance to extinction when marbles were withdrawn contingent on incorrect responses (W$\cancel{R}$) than when marbles were presented contingent on correct responses (R$\cancel{R}$), as was predicted from previous studies (Meyer & Seidman, 1961; Penney & Lupton, 1961). A significant interaction effect with age indicated that the facilitating effect of negative over positive feedback was greater for the young hyperactives than for the older hyperactives or the young and old normal boys.

In general, these studies using verbal or tangible reinforcers suggest that negative feedback acts to reduce impulsive responding and enables impulsive and hyperactive children to respond more like normal children under negative reinforcement responses, thus improving their performance on certain tasks.

There are very few studies investigating feedback in the CAL situation. Ball (1979) was concerned with feedback effects in a more difficult, abstract concept identification task adapted from the Reitan-Indiana Category Test (Reitan, 1959). This task included six subtests representing six different concepts and was computer administered under verbal feedback conditions of R$\cancel{R}$, W$\cancel{R}$, and RW. The hyperactive children did not perform as well as the matched normal control children on this task under any of the feedback conditions. Feedback effects were apparent only in the first three subtests; for both groups the most effective condition was

feedback for both correct and incorrect responses (RW). The fact
that RW was the most effective form of feedback was considered to
be due to the greater amount of information provided by the increased frequency of feedback. The general group differences between the hyperactives and normals were interpreted as an indication
of the attentional deficits of the hyperactive child preventing his
adequate functioning in a problem-solving situation. The lack of
differential feedback effectiveness does not support previous
research suggesting that negative feedback (W∅) acts to reduce impulsive responding and enables hyperactive children to perform like
normal children. It should be noted that studies supporting the
previous hypothesis have not looked at feedback effects in the CAL
situation. It may be that the visual presentation of a word on the
screen has informational value rather than acting as a motivational
factor.

PRESENT STUDY

    The following study was designed to clarify some of the issues
arising from the studies discussed above in the CAL situation.
Firstly, this study investigates further the differential effectiveness of the three feedback contingencies, R∅, W∅, and RW, in the CAL
learning situation. Secondly, the performance of learning disordered children was compared to that of hyperactive and normal children to determine whether they have greater difficulty on a variety
of learning tasks. Thirdly, three learning tasks were implemented
to represent three types of learning situations normally encountered
in the school setting. The major objective of this study is to
examine the most effective CAL methods of feedback for hyperactive
and learning disordered children with different types of learning
tasks.

    Teachers were given descriptive criteria for the preliminary
selection of hyperactive children in which the principal characteristics included high activity level, and impulsive, distractible
and emotionally labile behavior of a long-standing nature. Following this initial selection procedure, teachers were asked to
complete the Conners (1969, 1970) Teachers' Questionnaire (Conners
Preliminary School Report) for each of the children selected, and
those scoring 1.5 or greater on the hyperactivity index were included in the study.

    Similarly, teachers were provided with a description of a
learning disabled child, which primarily included average intelligence accompanied by poor academic performance. Subsequently, the
Myklebust Pupil Rating Scale was administered to these children.
Selection of subjects for the learning disabled group was made on
the basis of scoring less than 60 on this scale.

A control group of normal children was selected, matched for age, sex, grade and intellectual level with both the hyperactive and learning disabled groups. The normals were also required to score below 1.5 on the hyperactivity scale of the Conners Teacher Questionnaire and above 60 on the Myklebust Scale.

All children participating in the study were administered the Peabody Picture Vocabulary Test (PPVT) to provide an estimate of intellectual functioning. Only those children with PPVT IQ scores of 80 or above, with no known neurological damage and who were not taking psychotropic medication were included. Statistical analyses revealed no significant differences on the variables of age, sex, grade and IQ among the three groups.

The children were presented with the three automated learning tasks representing three types of task: paired-associate learning, simple discrimination learning, and abstract concept identification tasks. The learning tasks were administered via the screen of a remote computer terminal linked by telephone line to the main-frame computer at Carleton University. The order of task presentation was systematically varied, as was the feedback condition: Rb and Wb and RW. The task order and feedback condition were varied independently, with all children completing the three tasks.

Learning Tasks

In paired-associate learning, 10 pairs of stimulus items (one trial) were repeatedly presented to the child, until the child reached the criterion of performance or until eight trials had been completed. The ceiling criterion established was that of one trial in which eight of the 10 items was correctly paired. A percent score was then calculated based on the total number of correct responses on all trials, with all items in those trials above the ceiling scored as correct. The percent of items correct, including those above the ceiling criterion, was calculated on the basis of the total number of items.

The discrimination learning task involved the presentation of items, each item being one trial set of stimuli representative of the concept to be learned. Subsequent trials included stimuli representing the same concept. The measure of discrimination learning was that of trials to criterion, that is, the number of trials required to reach a ceiling criterion of performance, in this case nine correct responses in 10 consecutive trials, up to 100 items or trials. A percent score was then calculated. Thus, the percent of items correct, including those above the ceiling criterion, was calculated on the basis of the total number of items.

The abstract concept identification task involved the presentation of twenty items in each of six subtests which represented five
concepts and one memory subtest. The percent score was thus calculated based on the number of items correct as a percent of the total
number of items administered.

Results

The principal analytical technique was a three-way analysis of
variance of the percent correct scores. This analysis included the
three independent variables of feedback condition (R∅, W∅, and RW),
group (hyperactive, learning disabled and normal), and task
(paired-associates, discrimination learning and abstract concept
identification).

There were no significant main effects for the feedback
variable. Significant main effects for group and task were found.
These main effects indicate that the tasks were of different levels
of difficulty for all groups and that the groups performed differently on each of the tasks. Although there were no significant
interactions, there was a trend for the learning disabled children
to do least well on the paired-associates and discrimination learning task, and for the hyperactive children to do least well on the
abstract concept identification task.

Discussion

Feedback effects. In considering the differential effectiveness of feedback contingencies, it is important to distinguish
between the mode of presentation (verbal, tangible as compared to
CAL), and the type of child (normal, reflective-impulsive, hyperactive). As suggested in the introduction, feedback studies using
both verbal and tangible reinforcements with normal children have
provided relatively consistent agreement that the R∅ condition
results in poor learning. In addition, for normals, there is some
consensus that the inclusion of negative feedback contingencies,
either alone (W∅) or in combination with positive feedback (RW),
results in more effective learning. In the case of reflective-
impulsive children with tangible reinforcements, the W∅ condition
results in improved performance only for the impulsive group (e.g.,
Massari & Schack, 1972). For hyperactive children, with verbal
reinforcements, no differential effects of three types of feedback
were found (Firestone, 1974), while Cunningham and Knights (1978),
using tangible reinforcers, report a differential effect in favor
of the W∅ condition for younger hyperactives, but not for the older
group.

The two studies which examine the differential effectiveness
of visually presented feedback in a CAL situation are those of Ball

(1979) and the present study. In Ball's CAL study (1979), the differential effects of type of feedback were evident only in the first
three subtests of the abstract concept identification test and it
was the third subtest which primarily contributed to this significant difference. The most effective feedback contingency for this
subtest was the RW condition, but there were no differential effects
apparent between the hyperactive and normal children.

In the present CAL study, no differences were found as a
function of the R∅, W∅, and RW conditions, for any of the groups of
normal, hyperactive, or learning disabled children. It seems,
therefore, that in the CAL situation, where feedback is administered
in an automated form on the screen, it seems to be most appropriate
to apply an informational model in contrast to the motivational
model. In this particular instance, since detailed instructions
were given to the subjects concerning the meaning of the feedback,
including the blank condition, each of the three types of feedback
contained an equal amount of information. The proposed positive or
negative effect of different reinforcement contingencies appear to
have no motivational impact when administered by a computer.

Group effects. In general, the learning disabled children performed the worst and the normal control children performed the best
on the learning tasks presented. The inferiority of performance of
the learning disabled children was consistent for all dependent
measures and under all feedback conditions, with the exception of
the abstract concept identification task on which the hyperactive
children performed most poorly.

Since the groups were matched for age and IQ, the inferior
performance of the learning disabled group reflects the learning
difficulties they have in an academic setting. The poorer performance of the hyperactives on the abstract concept identification
task may be explained by their poor attention skills, preventing
them from developing effective problem-solving strategies.

Task effects. In general, the three tasks were different in
difficulty level for the three groups of children. As previously
described, the learning processes required for success were different for each of the tasks. The discrimination learning task was
the least difficult, followed by the abstract concept identification
task and finally the paired-associates task. This order of difficulty is a function of the length of the test, the level of ab-
straction, and verbal memory skills.

Those studies which have shown CAL to be particularly effective
have involved maximum feedback in drill and practice programs
(Knights & Hardy, 1978). In the present study, however, where new
learning is required, group differences remain.

The learning disabled children still had the most difficulty on tasks that required learning and memory skills, while the hyperactive children had most difficulty on the tasks requiring sustained attention skills.

Conclusions

In general, the effectiveness of feedback in the computer-aided learning situation appears to be different than when tangible or verbal reinforcements are used. The results of CAL studies support the informational model and show no real differences between the Rᴸ, Wᴸ, and RW conditions. In all of these three conditions the only change for the child is the nature of the word appearing on the screen. It is sensible to conclude that in this situation the child uses the information rather than reacting to the motivational aspects of the word presented.

The research on the type of task reveals relatively consistent results for learning disabled and hyperactive children. The learning disabled children still had the most difficulty on tasks that required learning and memory skills, while the hyperactive children had most difficulty on the tasks requiring sustained attention skills.

In conclusion, the present study does not provide any evidence for the greater effectiveness of CAL with exceptional children. In the situation where new learning is required, the hyperactive and learning disabled children do not perform as well as their normal peers.

References

Atkinson, R. C.   Teaching children to read using a computer.   American Psychologist, 1974, 29, 169-178.

Atkinson, R. C.   Computerized instruction and the learning process.   American Psychologist, 1968, 23, 225-229.

Babcock, L.   The effects of verbal reinforcement on the discrimination learning of middle and low-middle class children.   Unpublished Bachelor's Thesis, Carleton University, Ottawa, 1977.

Ball, C.   Differential effect of feedback on learning in hyperactive children.   Unpublished Master's Thesis, Carleton University, Ottawa, 1979.

Brackbill, Y., & O'Hara, J.   The relative effectiveness of reward and punishment for discrimination learning in children.   Journal of Comparative and Physiological Psychology, 1958, 51, 747-751.

Buchwald, A. M.   Experimental alterations in the effectiveness of verbal reinforcement combinations.   Journal of Experimental Psychology, 1959, 57, 351-361.

Buss, A. H., Braden, W., Orgel, A., & Buss, E. H.   Acquisitions and
    extinction with different verbal reinforcement conditions.
    Journal of Experimental Psychology, 1956, 52, 288-295.
Campbell, S. B., Douglas, V. I., & Morgenstern, G.   Cognitive styles
    in hyperactive children and the effect of methylphenidate.
    Journal of Child Psychology and Psychiatry, 1971, 12, 55-67.
Cohen, N. J., & Douglas, V. I.   The effect of reward on delayed
    reaction time and the orienting response in hyperactive and
    normal children. Unpublished Manuscript, McGill University,
    Montreal, Canada, 1971.
Conners, C. K.   Symptom patterns in hyperkinetic, neurotic and
    normal children. Child Development, 1970, 41, 667-682.
Conners, C. K.   A teacher rating scale for use in drug studies with
    children. American Journal of Psychiatry, 1969, 126, 152-156.
Cunningham, S. J., & Knights, R. M.   The performance of hyperactive
    and normal boys under differing reward and punishment schedules.
    Journal of Pediatric Psychology, 1978, 4, 195-201.
Curry, C.   The effects of verbal reinforcement combinations in
    learning in children. Journal of Experimental Psychology, 1960,
    59, 434.
Douglas, V. I.   Stop, look and listen: The problem of sustained
    attention and impulse control in hyperactive and normal children.
    Canadian Journal of Behavioral Science, 1972, 4, 259-281.
Fahlam, W. G.   CAI: A review of the literature. Canadian Forces
    Personnel Applied Research Unit, Report 75-7, October, 1965.
Feldman, D. H., & Sears, P. S.   Effects of computer-assisted
    instruction on children's behavior. Educational Technology,
    1970, 10, 11-14.
Fike, S. D.   Computer-assisted discrimination learning in retar-
    dates as a function of presentation of verbal feedback combina-
    tions. Unpublished Master's Thesis, Carleton University, Ottawa,
    1977.
Firestone, P.   The effects of reinforcement contingencies and
    caffeine on hyperactive children. Unpublished Doctoral Disser-
    tation, McGill University, Montreal, Canada, 1974.
Firestone, P., & Douglas, V. I.   The effects of verbal and material
    rewards and punishers on the performance of impulsive and
    reflective children. Unpublished Manuscript, McGill University,
    Montreal, Canada, 1974.
Fletcher, J. D., & Atkinson, R. C.   Evaluation of the Stanford CAL
    program in initial reading. Journal of Educational Psychology,
    1972, 63, 597-602.
Freibergs, V., & Douglas, V. I.   Concept learning in hyperactive
    and normal children. Journal of Abnormal Psychology, 1969, 74,
    388-395.
Gilman, D. A.   Comparison of several feedback methods for correct-
    ing errors by computer-assisted instruction. Journal of Educa-
    tional Psychology, 1969, 60, 503-508.

Gilman, D. A.   Feedback, prompting and overt correction procedures in nonbranching computer-assisted instruction programs. Journal of Educational Research, 1967, 9, 423-427.

Hartley, J. R., Lovell, K., & Sleeman, D. H.   Final Report to the Social Science Research Council. Leeds, England: University of Leeds, Research Council's Computer-Based Learning Project, December, 1972.

Hemry, F. P.   Effect of reinforcement conditions on a discrimination learning task for impulsive versus reflective children. Child Development, 1973, 44, 657-660.

Johnson, D. F., & Mihal, W. L.   Performance of blacks and white in computerized versus manual testing environments. American Psychologist, 1973, 28, 694-699.

Kagan, J.   Impulsive and reflective children: Significance of conceptual tempo. In J. D. Krumholtz (Ed.), Learning and the educational process. Chicago: Rand McNally, 1965.

Knights, R. M., & Hardy, M. I.   A Child-Computer-Teacher Assessment and Remedial Program for Child with Poor Reading Skills: Phase III. Research Bulletin No. 17, Carleton University, Ottawa, 1978.

Majer, K., Hansen, D., & Dick, W.   Note on effects of individualized verbal feedback on computer-assisted learning. Psychological Reports, 1971, 28, 217-218.

Massari, D. J., & Schack, M. L.   Discrimination learning by reflective and impulsive children as a function of reinforcement schedule. Developmental Psychology, 1972, 6, 183.

Meyer, W. J., & Offenbach, S. I.   Effectiveness of reward and punishment as a function of task complexity. Journal of Comparative and Physiological Psychology, 1962, 55, 532-534.

Meyer, W. J., & Seidman, S. B.   Relative effectiveness of different reinforcement combinations on concept learning of children at two developmental levels. Child Development, 1961, 32, 117-127.

Ochnocki, T. E., Cotter, P. D., & Miller, F. D.   Verbal reinforcement combinations, task complexity and overtraining on discrimination reversal learning in children. Journal of General Psychology, 1974, 90, 213-219.

Parry, P., & Douglas, V. I.   The effects of reward on the performance of hyperactive children. Paper presented at the Canadian Psychological Association, Vancouver, 1974.

Penney, R. K.   Effects of reward and punishment on children's orientation and discrimination learning. Journal of Experimental Psychology, 1967, 75, 140-142.

Penney, R. K., & Lupton, A. A.   Children's discrimination learning as a function of reward and punishment. Journal of Comparative and Physiological Psychology, 1961, 54, 449-451.

Reitan, R. M.   The effects of brain lesions on adaptive abilities in human beings. Indianapolis: Indiana University Medical Center, Mimeo, 1959.

Rockart, J. F.   Computers and the learning process.   Paper
    presented at the EDUCOM Conference, Toronto, Canada, 1974.
Schmeck, R. R., & Schmeck, E. L.   A comparative analysis of the
    effectiveness of feedback following errors and feedback following
    correct responses.   The Journal of Genetic Psychology,   1972, 87,
    219-223.
Seltzer, R. A.   Computer-assisted instruction: What it can do and
    cannot do.   American Psychologist,   1971, 26, 373-377.
Spence, J. T.   Verbal discrimination performance under different
    verbal reinforcement combinations.   Journal of Experimental
    Psychology,   1964, 67, 195-197.
Spence, J. T., & Segner, L. L.   Verbal versus nonverbal reinforce-
    ment combinations in the discrimination learning of middle and
    lower-class children.   Child Development,   1967, 38, 29-38.
Stevenson, H. W., Weir, M. W., & Zigler, E. F.   Discrimination
    learning as a function of motive incentive conditions.
    Psychological Reports,   1959, 5, 95-98.
Suppes, P., & Searle, B.   Computer teaches arithmetic.   School
    Review,   1971, 79, 213-225.
Tait, K., Hartley, J. R., & Anderson, R. C.   Feedback procedures
    in computer-assisted arithmetic instruction.   The British Journal
    of Educational Psychology,   1973, 43, 161-171.
Tindall, R. C., & Ratliff, R. G.   Interaction of reinforcement
    conditions and developmental level in a two-choice discrimination
    task with children.   Journal of Experimental Child Psychology,
    1974, 18, 183-189.
Whitehurst, G.   Discrmination learning as a function of reinforce-
    ment condition, task complexity and chronological age.   Journal
    of Experimental Child Psychology,   1969, 7, 314-324.
Witte, K. L., & Grossman, E. E.   The effects of reward and punish-
    ment upon children's attention, motivation and discrimination
    learning.   Child Development,   1971, 42, 537-542.

PART IV

Cognitive-Motor Development

Two major ideas occur over and over again in the chapters which
follow: It is difficult to separate cognitive and motor functions,
and the role of planning is evident in learning motor skills.
Whiting makes the point that the distinction between perceptual-
motor and cognitive processes is difficult to make. In promoting
his notion of the 'image of the act' he compels the reader to consider intentions, images, and schema as integral parts of learning
skilled-movement. Indeed, the representation of reality becomes
possible in a child through movement - a notion which is consistent
with the thinking of Piaget and Bruner. Whiting adds the names of
Bernstein and Konorski to this list of thinkers. Lest the reader
may conclude that Whiting is discussing something which has value
only in philosophy, we hasten to point out that his chapter has a
practical side to it - he has given some thought to procedures for
training skilled movement. Such procedures could be incorporated
into a training program for learning disabled children who have
awkward motor behavior. The latter is the subject matter of the
two chapters which follow.

In his chapter, Jack Keogh traces the strong link between
learning disability and perceptual-motor training programs which
are associated with Kephart and, more recently, with Delacato and
Ayres. He makes the interesting distinction between moving to
learn and learning to move. Since movement takes place in a social
environment he pays as much attention to personal-social factors in
learning to move as he does to perceptual cognitive factors, such
as plans or schema. If the reader thinks that the notion of
intelligence does not have a universal definition, then consider
defining clumsiness, or as Wall calls it, awkwardness in movement.
Keogh could not find any agreement in regard to the diagnoses of
clumsiness. However, from a practical point of view, Wall seems to
assume that the awkward child can be identified and even be
treated in a movement clinic. Wall, too, emphasizes the role of
plans and programs in motor behavior, and in fact achieves an integration of cognitive and motor processes by advocating the
use of a model which proceeds from an executive system

215

through motor planning to routine motor programs. All learning
disabled children are not clumsy, nor are all clumsy children learning disabled. It would be interesting to speculate if those who are
both learning disabled and clumsy, have a specific deficit in making
plans and executing a program.

IMAGE OF THE ACT

H. T. A. Whiting and B. den Brinker

Department of Psychology
Interfaculty of Physical Education
The Free University
Amsterdam, The Netherlands

PART I

A Theoretical Approach

In being asked to make a contribution to a conference on
'Learning Disabilities' from within the topic area 'Psychomotor
aspects,' we were confronted with three sources of constraint:

1. Whiting's (1980) recent statements about the limited appli-
cability of knowledge gained from the experimental skill laboratory
to what we will call the more meaningful 'messy world of real
affairs.'

2. The unclear relationship that may exist between the psycho-
motor literature and the problems of learning.

3. The extremely short time that was available between the
receipt of the invitation to make a contribution to the conference
and the necessity to prepare a paper having at least some relevance
to the conference theme. This you may appreciate, to people not
actually involved in the field of learning disabilities presented
major difficulties.

With regard to the latter constraint, it seems most appropriate
to take some of the ongoing theorizing with which we are involved
and to make an attempt to relate such information to the concerns of
those people more directly involved with learning disabilities. We
think that this can be done in a meaningful way. At the same time,

we would like the reader to appreciate that we are at an early stage both of our own conceptualization and research so that much of what follows is of a tentative nature.

Our first departure point, is to signal that the topic to be discussed belongs more appropriately under the label cognitive motor

Whiting's (1980) major criticism of much of the existing experi mental laboratory approaches to motor learning was couched in the following terms:

> ... the developmental progress of the action systems of
> man is towards increasingly complex voluntary motor co-
> ordinations dependent upon environmental information
> which becomes more and more divorced from primitive
> imposed stimuli. The developmental progress (?) of the
> experimental laboratory seems to be towards less and
> less complex coordinations based upon more and more
> primitive stimuli. Perhaps this mismatch alone,
> accounts for the lack of relevance of the one for the
> other.

It is proposed that recourse to a more cognitively oriented approach might do something to redress the balance and at the same time, handle the additional critique that while it is easier to theorize about and research into, skills in which the cognitive component is removed or reduced to a minimum, it is to be expected that such theories and related research would only find application where cognitive control is not critical, i.e. when man is performing at his most mechanistic.

Such cognitive orientations might also provide a useful impetus to the study of learning disability which in turn might lead to more meaningful diagnostic procedures. The long standing tendency to polarize the practical, doing, making side of man's action systems and the so-called logical, conceptual, thinking side - as Whiting (1979) has recently written - is unfortunate. While as Best (1979) points out, 'intelligent' does not entail 'intellectual' - since the latter involves a considerable degree of thinking involvement - it is nevertheless difficult to deny that movements in subserving overt actions, have a cognitive involvement, i.e. they are intelligently carried out. Moreover, concepts like 'action-oriented perception' and consideration of some aspects of motor learning as the 'cogni- tive learning of intentions' go some way towards questioning the meaningfulness of such a polarization.

Those who take an even more extreme viewpoint, question the possibility of separating out - other than in overly simplistic terms - movement and cognition. Weimer (1977) - for example - in a

review article entitled "Motor theories of the mind" introduces a
motor metatheory in which he asserts:

> ... there is no sharp separation between sensory and
> motor components of the nervous system which can be
> made on functional grounds and that the mental or
> cognitive realm is intrinsically motoric, like all
> the nervous system.

More recently, Chase and Chi (1980) point out the similarity between
perceptual-motor and cognitive skills, particularly with respect to
the improvement which takes place over long periods and to 'skill
specificity,' concluding that:

> There is not in fact, much theoretical justification
> for differentiating perceptual-motor and cognitive
> skills.

With such approaches, we are being asked at the one extreme to
consider those classically designated 'sensory' processes like perception, imagery, vision, memory etc. as motor processes (Weimer,
1977) or activities (Saugstad, 1977) and at the other, we are being
asked to think of motor learning in terms of cognitive parameters
like 'intention' and system modelling (Whiting, 1980). The essence
of such thinking - albeit, in the context of 'imagery' - is
expressed succinctly by Paivio (1975):

> ... imagery is a dynamic process more like an inner
> perceptual-motor system than a passive recorder of
> experience.

If these forms of categorization about which we are speaking
were only the focus of semantic argument, the issue would be less
important. But, insofar as they have implications for intervention
procedures, it is important that their similarities and differences
be pursued. This will not however be done at the present time,
rather we have set the stage for the central topic of this paper,
which might be aptly stated as the 'central representation of
actions/movements.'

Central Representation of Actions/Movements

Perhaps more than any other writer - and certainly much earlier
in time than most - the Russian eclectic neurophysiologist Bernstein
(1967) appreciated the links between cognition and movement and the
way in which a cognitive-motor system might develop:

Over the course of ontogenesis each encounter of a
particular individual with the surrounding environment, with conditions requiring the solution of a
motor problem, results in a development ... in its
nervous system of increasingly reliable and accurate
objective representations of the external world,
both in terms of the perception and comprehension
involved in meeting the situation and in terms of
projecting and controlling the realization of the
movements adequate to this situation. Each meaningful motor directive demands not an arbitrarily coded,
but an objective, quantitatively and qualitatively
reliable representation of the surrounding environment in the brain. Such an action is also an active
implement for the correct cognition of the surrounding
world. The achievement or failure of a solution to
every active motor problem encountered during life
leads to a progressive filtering and cross-indexing of
the evidence in the sensory syntheses mentioned above
and their components.

That there is less agreement over the form of such representations
will be clear from what follows, although it would probably be
agreed - with Paivio (1971) - that examples such as that of
Bernstein:

> ... suggest that our knowledge of the world is repre-
> sented mentally in terms of rather fine-grained
> analogue structures that are distinct from linguistic
> representations, although interfaced with them.

It would seem strange and, we suggest, probably misguided to suppose
that such a representative system did not have implications for
every facet of overt human action and that its degree of elaboration
and sophistication lies in the child's earlier exposure to an extensive range of motor problems in an increasing range of environmental
contexts. As the neuropsychologist Brown (1977) confirms from the
field of movement pathology:

> The central point to be made about these conditions,
> however, is not the fact that impairments of thought
> or affect occur; these are admittedly often minor
> features of the clinical picture. Rather, such im-
> pairments indicate that the movement disorder
> reflects a disruption in the development of action
> as part of, not extrinsic to, the rest of cognition;
> in other words, the substrate of the disorder is not
> a defective instrumentality played upon by more or
> less intact higher structures but is disruption at
> an early cognitive stage.

To us, the implication of this kind of thinking is that the ability
to fomulate 'adequate working models of the future' – a criterion
for successful adaptation – is dependent upon the degree of differ-
entiation and integration of information about past encounters and
present conditions. Understanding a problem, reflects the ability
of the problem-solver to construct an adequate representation of the
world. The ability to successfully behave skillfully in the everyday environment is dependent upon the coherency of existing cognitive motor systems and the extent to which they enable a person to
understand the requirements of a new situation and hence the nature
of a) the actions appropriate to that situation and b) the pattern
of external forces which it is necessary to overcome.

The 'Schema' Notion

        Amongst those approaches to the problem of the central repre-
sentation of actions/movements, that of the 'motor schema' has
probably received the most attention. In an extensive review
article on 'The schema notion in motor learning theory,' van Rossum
(1980) invokes Bruner, Goodnow, and Austin's (1956) distinction
between 'formation' and 'attainment' in order to critically evaluate
existing experimental findings about motor schema development.
Formation is the process of abstracting rules from specific environmental events, or more simply, the obtaining of rules. Attainment
on the other hand is the process of applying those rules in a particular situation, i.e. the use of existing rules. This distinction
is an important one. For example, it has relevance for Schmidt's
(1975, 1976) 'variability of practice' hypothesis referred to later,
since it would appear from the critique of van Rossum, that most of
the studies which purport to be on motor schema formation are in
fact on motor schema attainment. van Rossum further maintains, that
in these studies, insufficient attention has been given not only to
the differences between 'formation' and 'attainment' but also to the
consequences of defining the schema as a cognitive construct and to
the dimension on which training (or extra experience) should be
given.

'Image of the Act' of 'Image of Achievement?'

        Bearing these points of critique in mind, we would like now to
move on to two related concepts – 'image of the act' and 'image of
achievement.' As a starting point, we would like to take Pribram's
(1971) expressed disagreement with the interpretation of Bernstein
(1967) over the nature of the representations to be found in and
outside the motor cortex. Pribram maintains – contrary to
Bernstein's predictions – that it is not topological properties of
space which are represented in the motor cortex, but the forces
exciting muscle receptors. This 'image of achievement' – in
Pribram's (1971) words – comprises 'the learned anticipations of the

force and changes in force required to perform a task rather than an
abstract model of external space.' At the same time, Pribram is at
pains to point out, that his interpretation does not detract from
Bernstein's overall insight that 'properties of the environment' not
configurations of muscles and joints become cortically encoded.

In our opinion, the conceptualization of Pribram need not be
incongruent with that of Bernstein, if it is assumed that images
related to human overt actions can be conceived of at different
levels. With respect for example, to the formulating of actions, it
is normal - following Miller, Galanter, and Pribram (1960) - to
speak in terms of plans. A plan according to Shaffer (1980) is:

> ... an abstract homomorphism of the performance,
> representing its essential structure.

but, as Shaffer points out:

> The plan is executed in a continually renewing suc-
> cession of higher-order units by a motor program,
> which may construct one or more intermediate repre-
> sentations leading to output, adding the details
> necessary to specify the movement sequence. This
> idea of using a hierarchy of abstract representa-
> tions to construct performance from an intention is
> analogous to recent proposals in artificial intelli-
> gence for programs that solve problems ...

Eventually, all images of motor acts at whatever level - if they are
not to remain mere ideas - have to be translated into muscular activity, which, to be successful, must be finely tuned to the existing field of external forces to be overcome. Thus, Pribram's
'image of achievement' seems a necessary and plausible description.
Moreover, his interpretation of images using the holograph metaphor
goes some way to provide a solution for the problem of memory capacity. However, we would not wish to restrict our interpretations
solely to what is happening in the motor cortex. We believe that
what den Brinker (1979) chooses to call the 'image of the act' and
which he maintains exists at 'higher' levels than the motor cortex
(Bernstein also uses the term 'higher directional engram') is a
necessary construct to adequately account for motor learning.
'Images of the act' we maintain, exist at some intermediate level
between the abstract topological representation of space alluded to
by Bernstein (1967) and the more precise momentary 'image of
achievement' referred to by Pribram. Such images are a specification of the essential form of the movements - a qualitative representation - necessary to tackle a particular motor problem. They
are schema forms in the sense referred to by Gallistel (1980) in his
recently published book:

> ... the schema for a skilled movement contains no
> elements that specify which group of muscles are to
> be active, in what order, and for how long. That
> sort of detail is left to lower centres to fill in
> when the code is activated.

That these are sensible interpretations is apparent, since as
Bernstein (1967) stressed, the production of movements is non-univo-
cal, i.e. under differing circumstances, a movement of the same form
is generated by means of differing patterns of neuromuscular activity and different movements may be produced by similar patterns.
Thus, the 'image of the act' specifies the <u>form</u> of the movement
rather than the pattern of neuromuscular activity to be used in producing that form. This incidentally, is not to deny that the capacity for imagining appropriate forms is dependent upon the elaboration of the cognitive-motor system previously expounded.

The precise nature of such an image is open to question. We are
however attracted by the idea of a central code which specifies the
spatial and temporal patterning of the movements rather than the
signals required to generate the movements. Gallistel (1980) clearly
has similar ideas when he proposes - in a similar manner to that
earlier proposed by Pribram (1971) - a Fourier transform model of
motor schema comprising six-element engrams, each engram specifying
a fourier component of the movement. The six element engram represents period, amplitude, reference phase, together with the three
possible movement planes. Mathematically speaking, the engram of
the target's trajectory is a six-dimensional vector. The advantage
of this and other possible mathematical interpretations - in relation
to the ideas produced later in this chapter - is that apparently,
transposition of the movement can easily be obtained by mathematically describable variations of selected elements of the code. For
example in handwriting, letters can be made varying in size and
position without destroying their essential form. Tempo can also
be changed in piano playing without losing the essentials of the
melody.

A Multi-coding Working Model for Motor Skill Learning

A <u>working</u> model - Figure 1 - has been developed to handle the
problem of the translation of centrally represented images into
observable acts (den Brinker, 1979). It owes much to some of the
constructs outlined in Paivio's (1975) Dual Coding Process Theory as
well as the ideas already put forward. The mechanism necessary to
translate images into observable acts is conceived of - following
Brener (1974) as a Central Motor Controller (CMC). The motor cortex
is considered to form its central part. In addition to imagery
spaces for verbal representations, are spaces for visual perception,
visual imagery, proprioceptive imagery, auditory imagery etc. There
are connections between the different image spaces but no point-to-
point representation in one another. While an elaboration of the

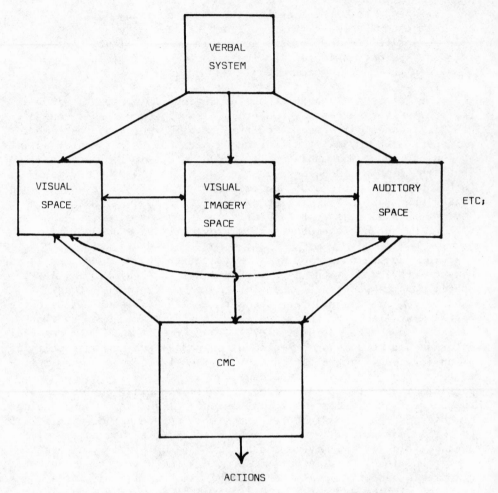

Figure 1.   A Multi-coding Model for Motor Skill Learning

structure of the CMC is not attempted, we are attracted by those
theories which propose overlapping functions for the latter stages
of the perceptual process and the early stages of movement production.   Konorski (1967) for example, elaborated on such an idea,
coining the term 'gnostic unit' to refer to the central representation of acts which once established could be utilized in a reverse
manner to produce movement.   More recently, Gallistel (1980)
proposes that perceiving a sinusiodal movement can culminate in the
creation of precisely those neural signals that are needed to direct
a corresponding voluntary movement.   Even more recently, Makay
(1981) in his fascinating paper on 'a general theory of serial order
in behavior' suggests on the basis of a series of transfer experiments:

> ... that input and output systems are closely related
> or identical at the conceptual level, rather than
> totally separate as is sometimes assumed.

Perhaps these are some interpretations of the integration of sensory and motor theories of the mind for which - amongst others - Weimer (1977) is striving.

The 'image of the act' has to be translated into an observable act via interaction with the Central Motor Controller (CMC) in which information is stored about the present and future fields of force which it is necessary to overcome (image of achievement). If an act is reproduced in another imagery space than that in which it was originated, less accurate results would be expected because of the non-existence of point-to-point representations of the differing imagery spaces in one another (Holding, 1966; Paivio, 1971; Scott-Kelso & Wallace, 1979). All the other image systems are considered to be subordinate to a verbal structure (or more generally, a symbolic structure (Bruner, 1973). In everyday language, we have many symbols which relate to acts in space - here/there, slow/fast, forceful/weak etc. The symbolic system is considered to have no direct access to the CMC. Symbolic acts have first to be translated in one or more of the non-verbal representations.

The Learning and Production of Overt Actions

In terms of the concepts introduced, learning disabilities may manifest themselves in two ways:

1. The 'image of the act' may measure up to some prescribed standard but the 'image of achievement' may be incorrect. The actor knew well the form of the movement but he was unable to operationalize it.

2. The 'image of achievement' may be an adequate translation of an inappropriate 'image of the act.'

Between these two extremes, all kinds of combinations of more or less appropriate 'images of the act' and 'images of achievement' are possible.

In most learning situations, inaccuracies in both images are likely to be encountered. It would seem to us, that the best way for an 'actor' to discover the properties of one or other of the images would be to keep the properties of the other images as stable as possible. Thus, in the formation of an 'image of the act,' we hypothesize that the learning situation should be so structured that the field of external forces to be overcome as well as the movement parameters amplitude, period and phase are kept relatively constant while the actor gives attention to establishing a reliable and appropriate 'image of the act.' It is likely that Bernstein (1967) had some such notion in mind, when he wrote:

> In exercises in sports and gymnastics, the motor struc-
> ture (referred to as style) is incorporated as an
> integral part in the meaningful aspect of the given
> problem. For this reason, it is one of the primary
> objects of the trainer to achieve as determinant a for-
> mulation and as rapid a stabilization of the motor
> structure as is possible of his pupil ...

In other words – and unlike the predictions from, for example, the 'variability of practice hypothesis' of Schmidt (1975) – the intro-duction of variations in environmental and movement parameter con-ditions should be postponed in the process of learning until an adequate 'image of the act' has been developed under one of the many conditions under which the act has eventually to be executed, i.e. 'the image of the act' has first to be developed as a holistic unit, a gestalt, before it can be manipulated to serve acts under changed conditions. It should be noted that this is not a plea for a form of training that 'grooves in' a stable 'image of the act' but is more in keeping with Bernstein's (1967) conception of training as being 'repetition without repetition' in the sense that the actor repeatedly attempts to solve the same motor problem by means which are adjusted on the basis of feedback. The updated 'image of the act' is distilled out of repeated attempts to solve the same problem

Once a usable 'image of the act' has been developed for say, a movement in two or three dimensional space, movements can be imag-ined that have the same general properties but which are for exam-ple, rotated to some degree or are larger or smaller in extent – providing that a change in the basic form is not also required. The insecurity about the field of external forces for the translated act remains and needs to be resolved in the course of further training.

The analysis put forward here, can perhaps be approached in another way. In perception studies, it is well appreciated that the discrimination of relevant aspects of a display can be retarded by the introduction of irrelevant information. Changing the environ-mental conditions in a way that demands extensive reformulation of the 'image of achievement' can result in a division of attention which may be detrimental to the establishment of a meaningful 'image of the act.' In a way, 'images of acts' are kinds of theory and like all good theories they need to be operationally tested and refined on the basis of feedback. Once formed, the 'image of the act' can be transformed without losing its general properties. Such procedures seem to be adopted in situations in which this results in time-saving or accuracy when compared with the recruiting of a new 'image of the act' involving an extension of the repertoire of the actor.

For the purpose of initial experimentation, we have chosen the skill of skiing, operationalized by the use of commercially available mechanical ski-simulators, together with a SELSPOT movement registration system. The ski-simulator consists of a platform that can ride over a pair of bowed metal runners (Figure 2) but which, because of the spring system under the platform, returns to the central resting position after being moved in a sideways direction. When the unloaded platform is disturbed by a horizontal force, it shows a damped oscillation of around 4 Hz.

Well trained people are able to make large and uniform movements in different tempos (maximum 1 Hz.) very similar to those which occur in slalom skiing. Untrained people find the skill very difficult and

Figure 2. Ski-simulator apparatus

generally only succeed in making small and irregular movements.
Such differences in movement of the platform can be picked up and
registered by means of the SELSPOT system. Figure 3 shows the output
traces of a well trained and an untrained person with respect to the
parameters amplitude and speed of movement.

A variety of techniques for the analysis of such data exist,
but these have seldom been used in psychological research on learning. For example, digital spectrum-analysis, auto and cross-
correlation. Pilot experimentation with a number of these measures
(den Brinker & van Hekken, 1981) suggested that for the parameter
'timing' an autocorrelation measure gave the most meaningful interpretation of learning effects and for the parameter 'fluency,' a
cross-correlation measure. In these pilot experiments, twelve naive
subjects (with respect to the apparatus and to slalom skiing)
trained for a period of four days in the making of gross, regular
movements. Such training comprised four periods of four minutes
interspersed with rest periods of three minutes per day. Before and
after each training period test measures were taken over periods of
30 seconds under two tempo conditions (36 or 42 complete movements
per minute). Subjects were required to make as <u>large</u> and as <u>uniform</u>
movements as possible in a prescribed tempo without falling off the
apparatus.

The results for the three parameters of interest are given in
Figures 4, 5, and 6. With respect to <u>amplitude</u> (Figure 4) subjects

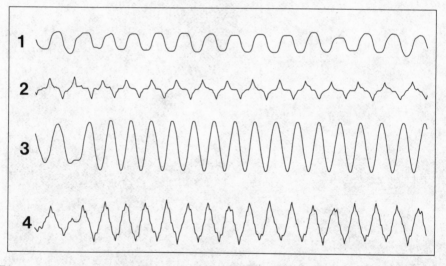

Figure 3. 1: position pattern of an untrained person
           2: speed pattern that is derived from position pattern 1
           3: position pattern of the same person after training
           4: speed pattern derived from 3

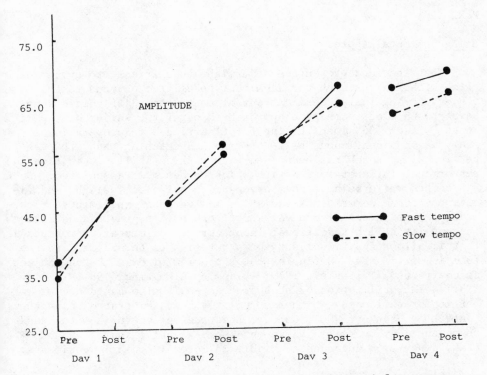

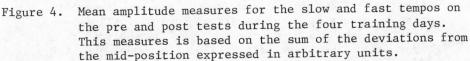

Figure 4.   Mean amplitude measures for the slow and fast tempos on
            the pre and post tests during the four training days.
            This measures is based on the sum of the deviations from
            the mid-position expressed in arbitrary units.

showed strong learning effects both with respect to the pre- and
post tests on a particular day as well as from day to day.   The
fall-off in gains towards the fourth day are reflected both in the
normal exponential growth curve to be expected in this kind of
learning as well as the restrictions imposed by the apparatus
design.   No differences in amplitude learning were shown between the
slow and fast tempos.

     With respect to the parameter of 'timing' (Figure 5), the most
noticeable and significant effect was the fall-off in performance
between pre- and post tests on the first day with subsequent im-
provements on each of the subsequent days.   It would appear that at
the beginning of the learning process, the subject is not able to
pay equal attention to improving his performance on all the parameters which are required of him.   He would seem to choose a strategy which in the first instance gives more attention to the amplitude
of the movement than the timing, i.e. there is a trade-off between
amplitude and timing.

PART II

An Experimental Approach

That the concepts introduced, might be difficult to operationalize within a research paradigm must be all too obvious to the
reader. Being fully cognizant ourselves of such difficulties, we
have chosen a more concrete approach to our first series of experiments. Since 1967 when Whiting first carried out an operational
analysis on a continuous ball-throwing and catching task, this experimental approach has been used with work on skill acquisition and
performance in laboratories with which he has been associated. Such
approaches while being laboratory based require only slight modifications to the normal task variables and various experimental treat-
ments can still be introduced. But, perhaps its greatest advantage
in comparison with traditional laboratory experimental approaches is
that it allows individual subject strategies to be utilized to the
full in solving the particular motor problem with which the subject
is confronted (Tyldesley, 1979; Tyldesley & Whiting, 1981; Whiting,
1981). The utilization of such procedures is also based upon a
particular philosophy which has recently been reiterated in another
context by Franks (1980), namely, that the outcome of a motor task
as measured by overall achievement (some global measure) cannot
stand alone as a dependent variable in motor learning experiments:

> Since the production of a movement involves the inter-
> play of several psychological and physiological sub-
> routines during a specified time period, the measure-
> ment used to describe motor performance should be both
> diverse and complete with regard to these complex
> processes.

Only by this kind of analysis - he suggests - can a comprehensive
description of the performance be built-up and a more integrated
view of the manner in which a subject acquires a movement sequence
be provided.

With these considerations in mind, we have decided to adopt a
paradigm involving the appraisal of cyclical movement patterns which
can be characterized by three measurements (see in this respect the
preliminary report of den Brinker and van Hekken, 1981):

1. The amplitude of the movement - the average absolute deviation from a middle position.

2. An auto-correlation coefficient as a measure of timing.

3. A cross-correlation coefficient as a measure of the fluency
or smoothness of the movement.

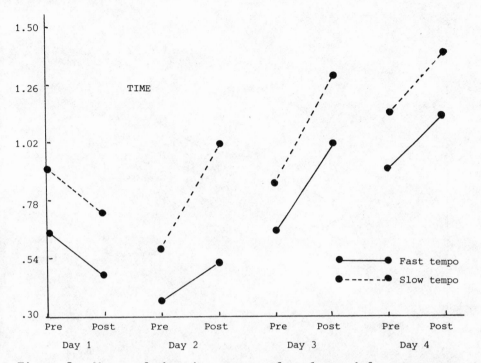

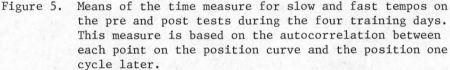

Figure 5.   Means of the time measure for slow and fast tempos on
            the pre and post tests during the four training days.
            This measure is based on the autocorrelation between
            each point on the position curve and the position one
            cycle later.

With respect to the parameter of 'fluency' (Figure 6), similar
learning effects are shown as for the parameter amplitude although
the fall-off in learning over days cannot now be attributed to the
design of the apparatus but is more determined by the exponential
nature of the learning curve.

These initial experiments confirmed that a more general measure
(the power of the frequency spectrum which provided information not
only about the tempo but also the amplitude of the movements) was
not the best measure to demonstrate learning effects.   But they confirmed that the more select parameters of amplitude, timing, and
fluency were of more explanatory value.   These will be used in
subsequent series of experiments which take as their departure point
the proposal that in early learning stages it is better to keep the
task requirements relatively stable rather than requiring training
on different measures of the same parameter.   In addition to the
three parameters of cyclic movements already named, our interest, as
has already been indicated, will focus on 'the external forces to be

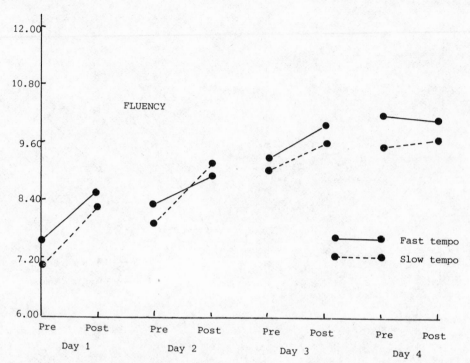

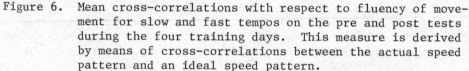

Figure 6.   Mean cross-correlations with respect to fluency of move-
            ment for slow and fast tempos on the pre and post tests
            during the four training days. This measure is derived
            by means of cross-correlations between the actual speed
            pattern and an ideal speed pattern.

overcome' - Pribram's (1971) 'image of achievement' - which we will
be able to manipulate by means of similar ski simulators utilizing
different tension spring systems.

    We appreciate that in these initial experiments our dependent
variable is a performance measure rather than images per se. While
reserving judgment on subsequent approaches until the results of the
first experiments are available, we are attracted by reinterpreta-
tions we have made of two traditional approaches to the laboratory
study of motor learning which, we suggest, could be re-adapted for
the study of imaging here discussed:

    1. Forced response 'guidance' learning. It will not have
escaped the notice of those actively involved in teaching, that very
often a teacher will physically put his pupil through a particular
movement pattern either because he wishes - along with Bernstein
(1967) - to establish an appropriate style from the beginning, or
because he considers that the learner has an inappropriate style and

he wishes to bring about changes. Now, while forced response
guidance has been shown to be an appropriate form of training
(Holding, 1966) in that it can produce enhanced performance, experi-
menters do not appear to have paid attention to the parameter for
which it would seem to be most appropriate, i.e. contributing to the
development of an appropriate 'image of the act' which may or may
not enhance performance of early trials following training. In a
recent series of studies, Hall (1980) uses forced-response guidance
in connection with imaging studies of movements, but is asking a
different question. He was able to show that:

> ... movement patterns can be scaled on imagery and
> recognition memory for movement patterns is a direct
> function of the imagery value associated with a
> pattern.

In a similar way, 'restriction' guidance in which the subject is
free to generate his own movement form but is prevented from moving
outside the confines of the required outcome may provide the possibility for operationalization of the 'image of achievement.'

2. Mental practice, i.e. whether skill can be facilitated by
mental or conceptual rehearsal of a behavior sequence in the absence
of muscular movements. The use of mental practice as an aid to
skill learning has had a chequered history with different studies
often producing conflicting results. Interpretations of facilitative
effects have included the fact that thinking about a movement
produces innervation of related muscles. But, once again, in as far
as subjects undergoing mental practice are often asked to imagine
themselves performing the movements, it would seem that its contribution towards the development of an appropriate 'image of the act'
might be a more meaningful interpretation. As Makay (1981) points
out:

> ... studies of mental practice have mainly been directed
> towards issues of proficiency as measured by the rate
> and efficiency of executing the component processes of
> sequential activities. And, unlike serial order, the
> problem of mental practice is not widely known: no great
> theorists such as Hull and Lashley have attempted to
> solve the problem and have failed.

CONCLUSION

In this paper, we have attempted to focus on a particular
cognitive aspect of skill acquisition. The theorizing is tentative
and exploratory, but we think may give considerable insight into the
way in which complex skills are acquired and hence possible sources

of learning disability. We regard it very much as an initial probe, being fully aware that most meaningful skills are acquired in a social context and that it is necessary to extend our thinking to encompass such a viewpoint, particularly with respect to the related problem of 'imitation.'

References

Bernstein, N. The coordination and regulation of movement. London: Pergamon, 1967.

Best, D. Philosophy and human movement. London: George, Allen & Unwin, 1979.

Brener, J. A general model of voluntary control applied to the phenomena of cardiovascular change. In P. A Obrist, A. H. Black, J. Brener, and L. V. Dicara (Eds.), Cardiovascular psychophysiology. Chicago: Aldine, 1974.

Brown, J. Mind, brain and consciousness. New York: Academic Press, 1977.

Bruner, J. S. Organization of early skilled action. Child Development, 1973, 44, 1-11.

Bruner, J. S., Goodnow, J. J., & Austin, G. A. A study of thinking. London: Wiley, 1956.

Chase, W. G., & Chi, M. T. H. Cognitive skill: Implications for spatial skill in large scale environments. In J. Harvey (Ed.), Cognition, social behavior and the environment. Potomac, M.D.: Erlbaum, 1980.

den Brinker, B. P. L. M. The influence of variations in training procedures on the learning of complex movement patterns. Unpublished paper, Department of Psychology, Interfaculty of Physical Education, The Free University, Amsterdam, 1979.

den Brinker, B. P. L. M., & van Hekken, M. F. The analysis of slalom ski movements on a ski simulator. Sport en Geneeskunde, 1981 (in press).

Franks, I. Unpublished Doctoral Dissertation. School of Physical Education, University of Alberta, Edmonton, Canada, 1980.

Gallistel, G. R. The organization of action - a new synthesis. Hillsdale, N.J.: Erlbaum, 1980.

Hall, C. R. Imagery for movement. Journal of Human Movement Studies, 1980, 6, 252-264.

Holding, D. H. Principles of training. London: Pergamon, 1966.

Konorski, J. Integrative activity of the brain. Chicago: University Press, 1967.

Makay, D. G. A general theory of serial order in behavior. Unpublished paper, Psychology Department, University of California at Los Angeles, 1980.

Miller, G. A., Galanter, E., & Pribram, K. H. Plans and the structure of behavior. New York: Holt, 1960.

Paivio, A. Neomentalism. Canadian Journal of Psychology, 1975, 29, 263-291.

Paivio, A.   Imagery and verbal processes.   New York: Holt, Rinehart
    & Winston, 1971.
Pribram, K. H.   Languages of the brain.   New Jersey: Prentice-Hall,
    1971.
Saugstad, P.   A theory of communication and use of language.
    Oxford: Global Book Resources, 1977.
Schmidt, R. A.   The schema as a solution to some persistent problems
    in motor learning theory.   In G. E. Stelmach (Ed.),   Motor
    control: issues and trends.   New York: Academic Press, 1976.
Schmidt, R. A.   A schema theory of discrete motor skill learning.
    Psychological Bulletin,   1975, 82, 225-260.
Scott-Kelso, J. A., & Wallace, S. A.   Conscious mechanisms in
    movement.   In G. E. Stelmach (Ed.),   Information processing in
    motor control and learning.   New York: Academic Press, 1978.
Shaffer, L. H.   Analyzing piano performance: a study of concert
    pianists.   In G. E. Stelmach and J. Requin (Eds.),   Tutorials in
    motor behavior.   Amsterdam, North-Holland, 1980.
Tyldesley, D. A.   Timing in motor skills.   Unpublished Doctoral
    Dissertation, Department of Physical Education, University of
    Leeds, 1979.
Tyldesley, D. A., & Whiting, H. T. A.   Sport psychology as a
    science.   In E. Geron (Ed.),   Introduction to sport psychology,
    1981 (in press).
van Rossum, J. H. A.   The schema notion in motor learning theory:
    some persistent problems in research.   Journal of Human Movement
    Studies,   1980, 6, 269-279.
Weimer, W. B.   A conceptual framework for cognitive psychology:
    Motor theories of the mind.   In R. Shaw and J. Bransford (Eds.),
    Perceiving, acting and knowing.   Hillsdale, N.J.: Erlbaum, 1977.
Whiting, H. T. A.   Skill in sport - a descriptive and prescriptive
    appraisal.   Proceedings of the 5th World Congress on Sport
    Psychology, Ottawa, Canada, 1981.
Whiting, H. T. A.   Dimensions of control in motor learning.   In
    G. E. Stelmach and J. Requin (Eds.),   Tutorials in motor behavior.
    Amsterdam: North-Holland, 1980.
Whiting, H. T. A.   Moving into action!   Inaugural lecture, Depart-
    ment of Psychology, Interfaculty of Physical Education, The Free
    University, Amsterdam, 1979.

THE STUDY OF MOVEMENT LEARNING DISABILITIES[1]

Jack Keogh

Department of Kinesiology
University of California, Los Angeles

Movement development has been an integral but confusing consideration in studying learning disabilities. Movement skills of children with particular learning problems have been measured; movement performance has been used to assess perceptual-cognitive and neurological functioning; moving has been used as a means to involve children in learning experiences. Two underlying propositions are that inadequacies in movement development are part of a direct, although involved, linkage to learning problems and moving provides an important, perhaps critical, modality for learning. The educational impact of the more than twenty years of advocacy for these propositions is that the movement development of children now receives more attention and moving is included in more ways in more educational programs. The scientific impact is negligible and often disruptive in terms of understanding learning disabilities.

The two general purposes of this paper are to (1) examine our current views on movement development as related to learning disabilities and (2) propose a major change in our research direction. The major shift in research direction is that we should focus directly on movement learning disabilities rather than continue to study movement development as a secondary and contributing condition in relation to other learning disabilities. Although this complicates the study of learning disabilities by adding yet another type of learning problem, a more fundamental understanding of movement

---

[1] Preparation of this chapter was supported in part by a training grant from the Office of Special Education (OEG007700997) and an award from a Biomedical Research Support Grant (USPHS-RR07009).

237

learning disabilities will provide a solid base when returning to
the study of movement development in relation to other learning
disabilities.

PERCEPTUAL-MOTOR TRAINING PROGRAMS

Movement development of children with learning disabilities has
been a concern at least as far back as the statements of Orton in
1937. The general propositions of Kephart (1960) mark the beginning
of a wider interest in what we now identify as perceptual-motor
training programs. These programs, which encompass much of our
thinking and research on this general topic, will be the focal point
of this brief review.

Two general and somewhat overlapping types of movement activities are used in perceptual-motor training programs. Movement
activities normally experienced by very young children (e.g.,
crawling, particular body positions) are used to help children go
through a normal progression of experiences and movement skill
changes. Children in these activities are being helped to retrace
their movement development with the idea of repairing and reorganizing neural structures and perceptual processes. Movement activities also are arranged as a means of directly affecting perceptual-
cognitive development and school achievements. These activities
range from arm movements during chalkboard activities for learning
about body and spatial directions to jumping from letter to letter
on the ground for learning spelling words.

The general premise shared by proponents of perceptual-motor
training programs is that movement development underlies and is
necessary to some extent for adequate perceptual-cognitive development, which then leads to academic achievement. Kephart (1960)
proposes that perceptual and cognitive development depends upon the
development of basic motor generalizations which are used to make
perceptual-motor matches. Both Delacato (1963) and Ayres (1972)
are concerned with an even more fundamental level of motor control.
The development and inhibition of primitive and basic reflexes are
viewed as providing neurosensory integration and organization which
are necessary for adequate perceptual-cognitive development.
Similar but more diffuse rationales are implied, if not stated, in
other perceptual-motor training programs and when investigators
measure movement development and performance of children with learning disabilities.

The general rationale for perceptual-motor training programs
involves a complex chain of events leading from movement development
through perceptual-cognitive development to school achievement. The
chain of events can be described in terms of three general concerns.

First, movement development is viewed as involving progressive changes which sometimes require specific types of movement experiences. Second, movement development is proposed as affecting perceptual-cognitive development in various ways. Third, perceptual-cognitive development is seen as affecting various aspects of school achievement. Deficiencies in the first will lead to problems in the second which will lead to problems in the third. Remediation strategies are based on what proponents see as important deficiencies and critical movement experiences.

A logical analysis of the general claims for perceptual-motor training programs indicates a number of difficulties in supporting such claims. Because the propositions are so broad in nature, it is not likely that a simple set of relationships can explain such an involved and lengthy chain of events. It seems more likely that individuals can compensate for particular deficiencies, thus increasing the complexity of the processes we are trying to understand. Another consideration is that learning disabilities appear to be multifaceted, whereas each perceptual-motor training program tends to focus on a limited aspect of movement development to explain most if not all learning problems.

There also is no compelling empirical support for the many propositions put forth by proponents of perceptual-motor training programs. The review by Myers and Hammill (1976) provides a comprehensive and critical coverage of research in this area. Very little research of significance has been produced beyond the date of their review (Cratty, 1981). They summarized published research in several ways to count the studies which reported positive, mixed or negative results relative to five areas of performance. Only one-fourth of the findings provide positive or mixed support even though a number of questionable studies were included in these counts.

A key proposition in the general perceptual-motor rationale is that individuals with learning disabilities should have movement development problems. The reverse should hold that individuals with movement development problems should have learning problems. Myers and Hammill (1976) identify a set of studies in which children with learning problems do not have the expected movement problems and children with movement problems achieve adequately in school. Most studies have been from a retrospective view of the general chain of events whereas a prospective view is needed to trace change in individuals.

Perhaps the most serious limitation in studying perceptual-motor training programs is the lack of systematic and unbiased programs of research. Published research has often been a single study by the investigator or the continuing work of those with vested interests in the training programs. Additionally, most

studies are poorly designed and conducted such that little confidence can be placed in the results.

Published research in this area has not been productive in
verifying or clarifying the relationships proposed for movement
development and learning disabilities. And there is little in this
literature to direct or redirect our thinking. A number of logical
considerations suggest that we need to rethink our position on
movement development and learning disabilities. Conceptual statements and empirical evidence are inadequate in all respects, adding
little to help us. As a starting point, we need to clarify the role
of movement in perceptual-motor training programs.

ROLE OF MOVEMENT

Learning to move and moving to learn are two quite different
ways to think about the role of movement in perceptual-motor
training programs (Keogh, 1978). The general premise has been that
problems in movement development must be remediated as the basis
for subsequent improvement in school achievement. Thus, an individual first must learn to move, granted that changes in movement
development are more than improvement in specific skills.

Another way to look at perceptual-motor training programs is
that they are movement experiences which provide opportunities to
gather information and engage in personal-social interactions.
Movement experiences in this view are a means to an end or an
environment for learning. Information is gathered in that moving
takes children to different locations, involves manipulation of
objects, and provides internal sensations and similar input. Information can be explored and elaborated through moving. This general
perspective includes personal-social interactions which are a
natural part of most movement activities. Movement situations can
be arranged to provide a wide range of personal-social information
and experiences.

Within this context of movement experiences as learning opportunities, children are moving to learn. Participation in movement
experiences to enhance learning has long been advocated as a form
of active learning (Cratty, 1971; Humphrey & Sullivan, 1970). Non-
movement experiences also provide learning opportunities but moving
is a powerful and useful means to achieve an end. Many children
enjoy moving, which gains their attention and involvement, and
movement experiences can be varied to provide a wide range of
learning opportunities.

The recognition of the two different roles of movement in
perceptual-motor training programs can help us separate our

educational and research concerns. Moving to learn can be an important educational experience with little need for empirical support
to justify the use of movement experiences as an instructional
environment. This is an educational decision which requires careful
thought to identify our educational concerns and values, and careful
planning to implement an educational program of movement experiences.
Moving to learn can be a useful and integral part of educational
programs without the need to invoke the explanations for learning
to move.

Much of the confusion in studying perceptual-motor training
programs has been the lack of clarity concerning the role of movement. Proponents claim and skeptics assume that learning to move
is the central issue. A careful analysis of what is done in most
perceptual-motor training programs and related research studies
suggests that children are moving to learn, regardless of the claims
and explanations offered. Also, the movement activities are quite
similar in many of the programs although the proponents claim different values for the movement experiences. If we are to continue
our research in pursuit of the long chain of events which begins
with learning to move and leads to school achievement, we must
separate and put aside the concerns which involve moving to learn.
The revised formulations must identify the links in the chain of
events and must specify relationships among these links in ways
which can be tested. Investigators then will be needed to conduct
systematic and long-term programs of research.

The previous paragraph provides my evaluation of what must be
done if the current direction of research is to be continued. The
general issue is important to many people but the chain of events
seems too long and complicated to be understood by such an approach.
Additionally, our thinking has been so global, imprecise and biased
that it seems unlikely that order and clarity will emerge from those
currently involved in this line of research.

A more fruitful research direction would be to concentrate on
problems in learning to move. As the first link in the chain, we
need to know more about learning to move before we can be concerned
about the impact of learning to move. A movement development perspective will be presented next as a basis for thinking about movement development problems.

DEVELOPMENT OF MOVEMENT CONTROL

The general flow of movement development has been traced by
observing changes in a variety of specific movements. The focus
early in life is on reflexes, postural control, locomotion and
manipulation as general types of movements to be developed. The

focus shifts after early childhood to general play-game skills
(e.g., jumping, throwing) and general abilities (e.g., coordination,
rhythm). Early movement development is measured in terms of mile-
stones or expected achievements which are assessed with developmental tests. Later movement development is more complex and confusing
with many different tests used to measure specific aspects.

Another way to conceptualize movement development is in terms
of movement control rather than types of movements (Keogh, 1981).
Are children developing control of their own body movements and are
they developing control of their movements in relation to external
movement demands? Control of movement-of-self is the first and
primary type of movement control in that children must be able to
control their own body movements before they can use their body
movements in relation to movement demands of the environment.
Establishing upright posture and thumbfinger opposition provide
control of one's body with little concern initially for how these
movements will be used.

Control of moving-with-others is a second type of control in
which a mover moves in relation to external demands. Movement now
becomes moving in that other persons and objects are moving and/or
the mover is moving. Moving changes spatial and timing relationships among self and others (persons and objects). Moving must be
planned and executed in reference to the initial and subsequent
location and movement of self and others. Catching a ball requires
the control of upright posture (in several positions) and appropriate arm-hand-finger movements. The control of posture and limb
movements in catching must be done at an appropriate location and
an appropriate time. Moving-with-others extends control of movement
to the use of movement in a moving environment.

Development of movement control in early years is primarily the
development of control of the neuromotor system to produce intended
movements in a reliable and consistent manner. A higher order of
development is the use of a movement repertoire to interact with
others. Control of movement-of-self, at an initial level of control, necessarily must preceed control of moving-with-others.
However, the two types of movement control must develop simultaneously and interactively. The availability of a larger and more
refined movement repertoire provides more and better choices when
moving-with-others. Moving-with-others should contribute to the
expansion and refinement of the movement repertoire as more sophisticated movements are experienced.

The distinction between movement-of-self and moving-with-others
is similar to Poulton's (1956) idea of closed and open skills. A
movement skill is closed when the mover can plan and execute a
movement in a static and predictable environment. A movement skill

is open when the mover must move in relation to a changing and less predictable environment. Few movements are completely open or closed, rather it is a continuum with a movement requiring more or less attention to environmental demands. Throwing a ball at a stationary target is essentially a closed skill in that the thrower can choose when to throw if the target is not moving. Throwing a ball at another child in a dodgeball game is more of an open skill because the other child is moving or likely will move.

An important consideration when thinking of movement develop-ment is the extent to which a child must deal with moving. Although all movements by definition involve moving, many movements can be accomplished with very little concern for changes in spatial rela-tionships and external timing demands. Movement in an open situa-tion involves processing of more information to analyze a movement problem and related sensory input, identify an adequate solution, and carry out the action plan. More information must be processed in less time; processing strategies must be improved to combine information and generate information beyond the information given; changes in movement strategies are needed to have more continuous rather than discrete responses. Moving-with-others increases the processing load and often requires a higher order of processing to plan and execute movements.

Movement in a closed situation allows the mover to determine when to initiate a predetermined movement. Movements may be a set of discrete or single movements rather than a continuous chain of responses. There are fewer problems of anticipation in that move-ment outcomes and effects are more predictable. Movement-of-self involves less of a processing load, both prior to and during the execution of the movement, and a lower order of processing.

Movement development has been characterized as the progressive development of movement control. Two types of movement control have been identified in relation to the amount and kind of process-demands. Throwing in various ways in different situations will be used to illustrate different levels of control within the two types of movement control.

Initial control of a throwing motion is control of movement-of-self, a closed skill. Initial control of the throwing motion, however general and crude, requires arm movement to bring the ball up or down and/or back, then forward to impart force to the ball. The release point and arm motion determine the direction the ball travels. Basic refinements of control will lead to appropriate wrist-finger flexion and other biomechanical changes to increase force and accuracy in throwing a ball in the general direction of a stationary target. A higher level of control will be seen as refinements or modifications of a particular throwing motion (overhand) and as throwing variations (underhand and sidearm). The achievement of a high level of performance (consistently throwing a dart in a bullseye) is indicative of a high level of control. Development of movement control is gaining initial control and achieving basic refinements and basic modifications. The achievement of a high level of control takes us beyond the concern of development of movement control.

Throwing often is used in more open situations to throw while moving or when others are moving. A throw in an open situation must be made quickly and at a particular time to a particular place. Also, formal and informal rules of play-game activities create open situations in that the thrower must throw to different places at different times depending upon the existing state of affairs. The important feature of throwing in an open movement situation is the need to have the throw coincide with the moving. Coincidence timing in various ways is an integral and demanding part of moving-with-others, whereas coincidence timing in relation to external demands is not an important concern in control of movement-of-self.

Playing catch with a partner is a movement situation that can range on a continuum from reasonably closed through very open. A throw to a stationary partner is a reasonably closed situation because the thrower selects when to throw and adjusts only to the distance and location of the partner. When the partner is running, the movement situation becomes more open. The thrower now must throw ahead of the partner to coincide the arrival of throw and partner at a predicted location. The movement situation becomes more unpredictable and more open when throwing to a partner who is being guarded by another player. The receiving partner also is in an open situation but in terms of catching rather than throwing the ball. Catching involves tracking an object and predicting a landing place while moving your body to be where the ball will land and organizing your arm(s) and hand(s) to contain the ball. When running to receive a throw, the catcher has few ways and little time to adjust to a ball in flight which is not accurately thrown to coincide with the path of the catcher. When guarded by another player, who also is trying to catch the throw, both thrower and partner must now include the moving of the other player in their movement plans.

Initial control and basic refinement of throwing in an open situation can be done only when the external demands are simple and few in number. Children then need to establish a small repertoire of simple and reliable "moving" throws which they can use in the many throwing situations they encounter. They must be able to throw ahead of their partner and with compensation for their own moving. They must have a basic sense of anticipation to plan and execute movements which will produce the necessary coincidence timing of self, objects and others.

The basic level of control to produce approximate coincidence timing must be modified and varied to produce better timing and movement combinations which can be applied to many situations. The thrower must deal with larger amounts of information in less time while combining and reorganizing information. Additionally, movement output becomes more complex and continuous. High level of performance in a game of keep away involves eluding another player while throwing to a partner and moving to receive a return throw from the partner, then continuously repeating the general exchange of roles.

Control of movement-of-self provides a reliable and consistent repertoire of movements. Control of moving-with-others is the use of the available movement repertoire to cope with the demands in an open or moving situation. Movement development problems now will be discussed in relation to the development of these two types of movement control.

MOVEMENT LEARNING DISABILITIES

Movement development problems must be viewed in relation to a full range of individual differences in movement development. At some arbitrary point on a continuum of movement performance, we become concerned that individuals below this point are not developing adequate control of their movements. Some children in this less adequate group have severe movement problems which are recognized as related to anatomical, neuromuscular or physiological limitations. The other children in this less adequate group have more subtle movement problems without a clear identification of underlying causalities.

The children with the more subtle movement problems and confusing diagnoses have a movement learning disability in the same sense that some problems in intellectual and personal-social development now are identified as learning disabilities. The definition of movement learning disability is a definition of exclusion, which is true of the definition of learning disability. Children with a movement learning disability are not severely limited in other aspects of development and are not diagnosed as having a related medical problem, but they do not move well and are not competent in movement activities. Children with a movement learning disability are a heterogeneous group in terms of defining characteristics as are children with learning disabilities. Movement learning disability is proposed as another type of learning disability in which children have a problem learning to move.

The general concerns for children identified as poor movers are that they do not move well and are not competent in movement activities. Not moving well is a qualitative observation which often is expressed by saying that a child is awkward or clumsy. This indicates a poor quality of control of movement-of-self. Children with severe movement control problems, such as cerebral palsy, are lacking initial control or have only minimal refinement. Clumsy-looking children have sufficient control to approximate the intended movement but they have not refined and modified the movement to appear more graceful.

The lack of competence in movement activities probably is a matter of not achieving in more open situations which involve timing and spatial relationships, and increased processing demands. Control of moving-with-others is not well developed in very noticeable cases but is not easy to observe in most cases. Inability to track moving objects or move in relation to moving objects is not difficult to observe. Moving with others in a complex and continuous situation is very difficult to observe and may be obscured by other behaviors.

Children with a movement learning disability can have varying combinations of movement inadequacies. Lack of competence may be observed only in particular movements activities and without any sign of poor quality of movement. Conversely, children may not look graceful while achieving adequately in particular activities, although poor control of movement-of-self probably is very limiting when moving-with-others.

Little is known about children with a movement learning disorder because movement problems have been studied secondarily for children identified as having other developmental problems. Additionally, movement performance has been measured in quite different ways and from quite different movement perspectives. Studies of clumsy children encompass the general sense of movement learning disorders and provide an empirical but cloudy starting point. It is important to recognize that there is no clear agreement on a definition of clumsy. Formal definitions range from more functional statements about maladaptive behavior (Morris & Whiting, 1971) to medical descriptions of developmental apraxia and agnosia (Gubbay, 1975). Children identified as clumsy presumably represent both the qualitative sense of not moving well and the functional sense of not competent in movement activities.

Samples of clumsy children have been drawn primarily from problem-presenting children in clinical and special school settings. Only two published reports have been found in which clumsy children were identified in a normal school sample. Gubbay (1975) has conducted the most extensive study of clumsy children, including 21 case studies in one clinical sample, 19 children with a comprehensive battery of psychological and medical tests, and a screening of

922 school children on a variety of tests. Keogh, Sugden, Reynard, and Calkins (1979) included a movement skills test and ratings of physical education specialists and classroom teachers, with 41 boys rated over a two-year school period with different teachers.

There is very little agreement within and across studies in the identification of clumsy children. No single measure of clumsiness has been found that will identify even one-half of the children in a group of clumsy children selected by several measures (Gubbay, 1975). A change in observers may change the identification, as happened when kindergarten teachers identified a different group of boys as clumsy than did first-grade teachers who had the same boys a year later (Keogh et al., 1979). Whiting, Clark, and Morris (1969) found the same lack of agreement when comparing medical diagnoses and performance on the Stott Test, but each procedure was useful in identifying problems not identified by the other assessment procedure.

Children identified as clumsy seem to be a heterogeneous group, even within one sample, in terms of defining characteristics. This is consistent with other learning disabilities. A more precise definition of clumsy is needed with a multiple measurement approach in identifying individuals as clumsy.

Incidence of clumsiness is difficult to estimate because identification procedures have not been comparable. A general estimate of incidence gathered from several studies is 5% or more for boys (Keogh et al., 1979). Approximately three times as many boys as girls have been identified as clumsy, except Gubbay (1975) reported no incidence difference for boys and girls. The larger proportion of boys identified as clumsy also is consistent with the finding for other learning disabilities. More attention within the study of movement learning disabilities should be focused upon problems of not moving adequately rather than on the multitude of secondary characteristics. Keogh et al. (1979) suggested what might be possible movement control problems and related behavioral problems. Kindergarten teachers noted more problems for their children in moving situations, particularly that children appeared confused, hesitant or lost. These behaviors need to be related back to specific movement control difficulties. Some of these children also were seen by their teachers as moving too much in terms of being impulsive, overenthusiastic and disruptive. Personal-social behaviors such as these may confuse our assessment of clumsiness and may even be used by children to draw attention from or conceal their clumsiness. Personal-social behaviors are an integral part of competence in movement situations and must be a part of future studies of clumsy children.

The findings and ideas about clumsy children provide our best sense of what is a movement learning disability. The lack of competence in movement situations appears to be real but confusing as well as difficult to define and identify. The major concern of poor quality of movement and lack of movement competence are overlaid and entangled with personal-social behaviors. Movement learning disability appears to have the same multifaceted and complex problems as are found in other learning disorders.

The general diagram in Figure 1, will be used to suggest three major explanations for movement development problems. The production of movement is represented as the result of processing information within a personal-social surround. Sensory input is gathered from internal and external sources. Perceptual-cognitive systems interpret input and formulate action decisions and plans. Neuromotor systems then produce movement which we observe and measure as movement performance and related outcomes. Movement occurs in a personal-social surround which provides information to be processed in planning and producing movement. The description thus far indicates what is involved in the production of a single, simple movement. A more complete picture must include movement sequences and continuity, and the interactions of the systems, surround and outcomes. Within this complex and dynamic sense of movement, three major sources of movement development problems are the neuromotor systems, sensory-perceptual-cognitive systems and personal-social surround.

Minor problems in the neuromotor systems may produce mild movement development problems which cannot be diagnosed medically and are observed as clumsy looking. The sensory-perceptual-cognitive systems can have similar minor malfunctions. Neuromotor problems are more likely to affect the control of movement-of-self, whereas sensory-perceptual-cognitive problems are more likely to limit control of moving-with-others. Problems in coping with the personal-social surround could affect both aspects of movement control, probably in an inconsistent manner. Children may not be

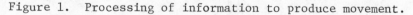

Figure 1.   Processing of information to produce movement.

comfortable with the personal-social demands of movement situations and the sensations of moving, including the possibility of pain and physical harm (Griffin & Keogh, in press). These children likely will avoid movement situations and will not perform well even if their neuromotor and sensory-perceptual-cognitive systems are adequately developed. Our observations of these children will be inconsistent, depending upon which movement situations they can and cannot handle and when we observe them.

Our educational assessment of children with a movement learning disability should begin with their personal-social comfort in movement situations. Next would be an assessment of their ability to cope with moving followed by an assessment of their ability to control their movements. Our selection of subjects for experimental purposes should follow a similar sequence to identify groups which are more homogeneous in terms of movement development problems. We currently lack the measurement tools to assess children or select subjects in the manner suggested, but we can construct such measurement tools when we decide they are needed.

Interactions among these three problem sources should be the rule rather than the exception which compounds our methodological and theoretical problems. Returning to the initial discussion of perceptual-motor training programs, the same problem exists for these programs in that singular explanations are offered for what must be multivariate problems. The proposed shift in research direction would take us away from the chain of events in the perceptual-motor explanations to look directly and only at movement development problems. Additionally, three sources of movement development problems are suggested to sharpen our focus while expanding our view of movement development. The interaction of the three problem sources and others which may be identified is something to be studied after more is known about each of the problem sources.

CONCLUDING COMMENTS

The basic thrust of this paper is that we need to study movement learning disabilities as a primary learning problem rather than as a secondary concern in relation to other learning problems. A general perspective of movement development was presented as a lead to identifying three major sources of movement learning problems. The three problem sources suggest what may be the nature of movement learning disabilities. The three problem sources can be viewed as two types of problems, one which directly affects the production of movement and one which indirectly affects movement by limiting or dampening the systems which produce movement. The neuromotor and sensory-perceptual-cognitive systems are directly involved in the production of movement whereas the personal-social surround indirectly affects the production of movement by providing information to be processed through the internal systems. The personal-social surround overlays and constrains the movement producing systems, and very likely is the major consideration for children who are not comfortable with movement. The important point is that movement situations are not positive for all children which must be considered when placing a child in a movement program to enhance learning and when selecting a child as an experimental subject.

The lack of theory in movement development is the most serious limitation in studying movement learning problems. A more careful description of movement learning problems in the context of important movement situations is one line of research which may help us organize some preliminary theoretical perspectives. And we must be careful not to be drawn back to the seductive lure of moving to learn rather than keeping our focus on learning to move.

References

Ayres, A. J.   Sensory integration and learning disorders. Los Angeles: Western Psychological Services, 1972.

Cratty, B. J.   Sensory-motor and perceptual-motor theories and practices: An overview and evaluation. In R. Walk and H. Pick (Eds.), Intersensory perception and sensory interaction. New York: Plenum, 1981.

Cratty, B. J.   Active learning: Games to enhance academic abilities. Englewood Cliffs, N.J.: Prentice-Hall, 1971.

Delacato, C. H.   The diagnosis and treatment of speech and reading problems. Springfield, Ill.: Charles C. Thomas, 1963.

Griffin, N. S., & Keogh, J. F.   A model for movement confidence. In J. E. Clark and J. A. S. Kelso (Eds.), The development of movement control and coordination. New York: Wiley, in press.

Gubbay, S. S.   The clumsy child. Philadelphia: Saunders, 1975.

Humphrey, J. H., & Sullivan, D. D.   Teaching slow learners through active games. Springfield, Ill.: Charles C. Thomas, 1970.

Keogh, J. F.   A movement development framework and a perceptual-cognitive perspective. In G. A. Brooks (Ed.), Perspectives on the academic discipline of physical education. Champaign, Ill.: Human Kinetics, 1981.

Keogh, J. F.   Movement outcomes as conceptual guidelines in the perceptual-motor maze. Journal of Special Education, 1978, 12, 321-330.

Keogh, J. F., Sugden, D. A., Reynard, C. L., & Calkins, J. A.   Identification of clumsy children: Comparisons and comments. Journal of Human Movement Studies, 1979, 5, 32-41.

Kephart, N. C.   The slow learner in the classroom. Columbus, Ohio: Charles E. Merrill, 1960.

Morris, P. R., & Whiting, H. T. A.   Motor impairment and compensa-
    tory education. Philadelphia: Lea and Febiger, 1971.
Myers, P. A. I., & Hammill, D. D.   Methods for learning disorders
    (2nd Ed.). New York: Wiley, 1976.
Orton, S.   Reading, writing and speech problems in children.   New
    York: Norton, 1937.
Poulton, E. C.   On prediction of skilled movements. Psychological
    Bulletin, 1956, 54, 467-478.
Whiting, H. T. A., Clarke, T. A., & Morris, P. R.   A clinical
    validation of the Stott Test of Motor Impairment. British
    Journal of Social and Clinical Psychology, 1969, 8, 270-274.

PHYSICALLY AWKWARD CHILDREN: A MOTOR DEVELOPMENT PERSPECTIVE[1]

A. E. Wall

Department of Physical Education
University of Alberta
Edmonton, Canada

At the present time in North American culture, proficiency in sport is highly valued. The financial and social benefits that are provided to professional athletes, and in some cases to amateur athletes, are clear manifestations of this fact. During the childhood years, proficiency in play and games is an important factor in the social and emotional development of children. Most children develop competency in movement skills with relative ease; others who become highly proficient do so from greater interest and practice. This chapter is concerned with children who do not develop adequate proficiency in movement skills. The first section of the chapter provides a definition of physically awkward children and describes the awkwardness syndrome. The second section reviews some of the key stages in the development of cognitive-motor skill and discusses some of the factors related to the acquisition of skill. The third section discusses individual differences in motor skill in relation to culturally-normative increases in motor task demands. The concluding section describes a clinical program for movement learning disabled children at the University of Alberta.

---

[1] I wish to thank Jane Taylor, Betsy Terry, and John Hay for their contributions to the writing of this paper. The financial support of the Alberta Recreation, Parks and Wildlife Foundation is appreciated.

THE AWKWARDNESS SYNDROME

Torgeson (1980) suggests that one of the most important problems in the field of learning disabilities is that of definition.
Central to his concern is the fact that the term "learning disabilities" refers to a large, heterogeneous group of children with very
different capabilities, difficulties and needs. He notes that if
learning disabled children are viewed as a homogeneous group then
the development of an adequate theoretical framework for their
identification and remediation will be severely hampered. We know
that a significant group of children, perhaps as much as six percent
of school children, experience serious difficulties in learning
motor skills (Gubbay, 1975; Keogh, 1968). We contend that these
children are a definite sub-group of the total population of learning disabled children. Unfortunately, all too often in the professional literature on learning disabilities, a rather narrow definition of motor proficiency is used. Therefore, movement learning
difficulties are often defined simply in terms of fine-motor behaviors that are "the means for representing what has been per-
ceived" (Keogh, 1978). Fine-motor difficulties are part of the
movement learning disabled child's problems; however, a more precise definition of the concept of physical awkwardness is required.
The following definition of physical awkwardness attempts to broaden
and sharpen the meaning of movement learning difficulties.

Physically awkward children are children without known neuromuscular problems who fail to perform culturally-normative motor
skills with acceptable proficiency. As with all definitions of
extremely complex phenomena, we need to clarify a number of ideas
that are included in this definition. Culturally-normative physical
skills are skills that are generally used within a specific culture
by a large majority of people. Skills such as running, jumping, and
climbing are culturally-normative in many environments, whereas
skills like catching a baseball, hitting a cricket ball, and high-
kicking a stuffed seal skin are identified with other cultural
environments. In North America, the skills of catching, throwing,
kicking and hitting a ball, swimming, and in some instances skating
and skipping are physical skills that are widely used in play,
games, and sports.

As Elliott and Connolly (1974, p. 135) note, proficiency in
skills refers to "an ability to achieve defined goals with an
efficiency beyond that of the inexperienced person." Proficiency
in skill is characterized by purposeful, planned, accurate, and
precise behavior. Unfortunately, acceptable proficiency is not so
readily defined. To a large extent, the band-width of acceptable
proficiency varies with the age, sex, and socio-cultural environment of the person. Even more so, the performance expectations of
significant others such as siblings, parents, teachers, and peers
certainly influence the standards of acceptable performance.

For a number of years, concerned educators have recognized that some children do not perform culturally-normative motor skills with acceptable proficiency. We contend that these children exhibit a cluster of behaviors which stem from their inadequate motor performance. Collectively these behaviors can be labeled an "awkwardness syndrome." Often, the most significant feature of the syndrome is the lack of motor skill exhibited by the child in play and games environments. In addition, the child may experience difficulty with such fine motor skills as using a pencil, cutting with scissors, and tying shoelaces. In more severe cases, awkwardness in everyday movement competencies such as running or carrying objects may result in ridicule by other children (Gordon & McKinlay, 1980).

As Whiting, Clarke, and Morris (1969) and others note, the awkward child may be excluded from social play and game situations which results in withdrawal from other group situations. Physical awkwardness is open to evaluation by others in a group. In contrast, a child with reading difficulties can attempt to hide this fact in a number of ways. However, if a child fumbles a ball thrown by a playmate, it is obvious to everyone involved. The negative reactions that the child receives may force him or her to withdraw from other group experiences and discourage involvement in play and game situations. In time, the child's peers label him or her as clumsy and exclude him or her from group play situations. Ultimately, the child's lack of motor skill, minimal enjoyment in physical activity and social difficulties within play situations combine to create a disinterest in physical activity and a corresponding low level of physical fitness.

The syndrome we have described is a serious one. The children that we see in our clinical programs certainly need to be identified early, require careful professional evaluation, and deserve the prescription of sensible remedial strategies.

DEVELOPMENT OF COGNITIVE-MOTOR SKILL

This section presents a brief review of some of the major factors underlying "normal" motor development. An understanding of these factors may prove helpful when discussing key issues related to the identification, evaluation, and remediation of physically awkward children.

In recent years, we have come to realize that motor development results from the interaction of genetic and experiential factors. A number of researchers provide evidence that supports Bouchard's (1976) recent conclusion that

... genes play a critical role in the unfolding of
normal motor behavior, in the achievement of high
motor performance, and in the maintenance of a given
level of fitness.

Although research has emphasized the interaction between nature and
nurture, the evidence is clear that the potential of an individual
is largely determined by one's genetic endowment. If potential is
set by genetic endowment, it is even more clear that reaching one's
potential is determined by a host of environmental factors. The
nutritional status of the mother and the adequacy of early nutrition
are of crucial importance to neuromuscular development. The
patterns of parental care are also key factors related to the development of movement competencies. The degree to which a child
develops certain motor skills will depend on the personal interests
and capabilities of family members, especially the parents or
primary caregivers. Furthermore, the interests of the family in
play and games will be largely determined by the past experiences of
the family members as well as the socio-cultural environment in
which they live.

The fundamental purpose of motor development is the acquisition
of motor skills that allow a person to efficiently meet the increasing task demands of his or her environment. To a large extent, the
basic motor skills of sucking, grasping, sitting, walking, and
running are species-specific skills that every person must learn if
he or she is to independently cope with life's demands. Considerable debate has centered on the degree to which these fundamental
skills arise from predetermined primitive reflexes, as Zelazo (1976)
suggests, or whether they emerge from instrumental approach and
avoidance activities in neonates (McDonnell, 1979). However, the
fact remains that these early motor skills are characterized by
purposeful behavior that becomes more accurate and precise with
maturity and practice.

For the purpose of this discussion, it is important to examine
the changes in task demands as a child matures. To do so, we must
consider what is meant by task demands. Higgins (1977) suggests
that coordinated, effective, goal-directed movements are possible
only because of the predictable features within an environment.
These predictable features can be classified on the basis of both
spatial and temporal task demands. The spatial elements of an
environment are concerned with the relative position of objects in
three-dimensional space; whereas, the temporal elements depend on
the sequential relationships between events and objects in space.

Increases in task difficulty depend on the spatial and temporal
characteristics of a particular environment. The concept of information load reflects the degree of uncertainty associated with a

particular performance environment. When a child reaches for a
stationary object such as a rattle, the spatial and temporal demands
are relatively low; in contrast, when an adolescent attempts to hit
a curve ball in a baseball game, the information load is much
greater due to the relatively unpredictable path that the ball will
take. A competitor in a professional tennis match must contend with
very high information loads due to the time constraints under which
the game is played and the skill of his or her opponent in masking
where and when the ball will be returned in a particular sequence of
play.

A central contention of this paper is that with increasing age,
the task demands made of a person will increase depending on the
socio-cultural environment of that person. During infancy, the
young child learns such skills as grasping, reaching, sitting,
standing, and walking. The developmental sequences of Bayley (1935)
and Gesell and Armatruda (1941) provide empirical support for these
observations. As the child grows older and enters what Bruner
(1973) has called the "mastery stage of play," the child learns to
run, climb up and down stairs, jump from and onto various objects,
tricycle, throw, kick, and strike objects. Up until this point in
the developmental progression, the central feature of these skills
is the organization of movement sequences that respond, in most
cases, to relatively low spatial and temporal information loads.
Therefore, the amount of prediction that is required prior to the
initiation of a response is relatively low.

Jack Keogh (1975) suggests that this period of motor skill
development might be labeled the development of movement consistency,
that is, the development of a repertoire of movement skills that are
characterized by the efficient patterning and ordering of movements
to solve everyday living problems in an appropriate and reliable
way. Furthermore, he suggests that as the child grows older, the
spatial and temporal demands of tasks increase and the resulting
information loads require the ability to predict and use flexibly
the skills that are learned during the movement consistency phase
of motor development.

Once the child reaches school age, the task demands with which
he or she must cope rapidly increases. Bouncing, catching, hitting,
and kicking are skills that are characterized by increases in
spatial and temporal uncertainty that demand the use of prediction
and other strategic behaviors. From ages 6 to 11 years, children
develop cognitive-motor skills that allow them to participate in
favorite playground games like tetherball and dodgeball or competitive dual and team games such as tennis, hockey, and soccer. In
fact, the work of Sutton-Smith (1961) in North America and Lindsay
and Palmer (1981) in Australia, on the favorite games of children
during the school age years support Keogh's (1977) observation that

the focus of motor development changes from the need for movement consistency to what he has called <u>movement constancy</u>. Movement constancy is characterized by the flexible use of movement consistencies in a variety of movement situations. Flexibility of movement is required due to the changes in task demands that are imposed by a variety of sources in a child's environment. The differences in skill required in simple playground games such as tag and hopscotch in contrast with touch football and ice hockey games support the above observations. The players in these latter games must contend with much greater spatial-temporal information loads due to the changes in position of the ball or puck and the time constraints within competitive team situations.

For a child to play effectively in these sport situations, he or she must be able to interpret such cues as the speed of a ball, where an opponent is likely to move, when an appropriate shot should or will be taken so that the amount of time in selecting a response will be cut down. In essence, the skillful player is better able to predict what will happen and when it will occur so that he or she can respond quickly and accurately. In order to become skillful the child must progress through the movement constancy and movement consistency phases of motor development at an age-appropriate pace. Some of the key factors underlying the acquisition of skills are discussed in the next section of this chapter.

PHYSICAL AWKWARDNESS AND THE ACQUISITION OF COGNITIVE-MOTOR SKILLS

In recent years, an especially effective view of the acquisition of motor skills has emerged from research based on information processing models. Researchers such as Fitts (1967), Whiting (1969), Welford (1968), Marteniuk (1974) and Schmidt (1976) describe the following processes in the information processing chain. The reception and perception of input stimuli, the organized storage of this information, the generation of appropriate decisions based on the use of suitable strategies, followed by the planning and efficient execution of movement responses that elicit feedback through the various loops which control and define the goal-directed behavior over trials.

Glencross (1978) extends the above ideas and postulates the existence of a hierarchical structure within given skills and relates it to the notion of plans. He suggests that there are four distinct levels of planning and that the capacity demands in terms of information processed by the performer decrease from the highest executive system level through the general plan level and motor plan level, to the routinized motor programs level. He purports that the executive system level is concerned with the desired outcomes, objectives and goals of skilled behavior; furthermore, this system is relatively "slow acting" and depends highly on intrinsic and extrinsic feedback which makes it a high capacity system in information processing terms. The general plan level is the motor plan control system which establishes the general features of a particular response type such as running, throwing, or kicking actions. Again, this system depends on feedback and is relatively high in capacity demands. Glencross (1978) views motor programs as "constructions of specific and detailed patterns of action which are superimposed upon the motor plan" (p. 83). At this stage, details are added to the general plan of a response making it relatively automatic or open-loop in nature; that is, it is less dependent on feedback, so capacity demands are smaller. Finally, the lower level motor programs are the subroutines that make up the motor plan. Throughout the four stages, Glencross (1978) contends that there are feedback loops that allow the performer to amend or correct his or her performance which helps explain the adaptability of skillful actions. However, the central feature of his model of skilled performance is the hierarchical nature of the levels, with higher order levels delegating control to lower levels that permit the performer to optimally meet the changing task demands of his or her environment.

    Glencross (1978) contends that plans at all four levels are continually modified through practice. He notes that during the early phases of skill learning the performer uses executive plans to monitor the various lower-level motor plans within a response. However, with practice, the performer can modularize "predictable sequences of actions and form larger units of action under motor program control" (p. 81). The skilled performer can execute relatively complex response actions with minimal sensory feedback; in contrast, the unskilled performer must use substantial amounts of information processing capacity to monitor the execution of a response. Therefore, the skilled performer who can relegate the control of a response to lower-level motor programs will be able to more readily handle tasks with greater spatial-temporal uncertainty. Furthermore, as responses become modularized, under the direction of lower level control systems, the skillful child is able to repeat movements with a greater degree of precision. The increased predictability of response production reduces the information with which the child has to cope, and therefore, he or she has more available capacity to process environmental information. Unfortunately, physically awkward children often have response execution difficulties that prevent them from using lower-level motor control plans. However, they are expected to improve their skill level through practice in age-appropriate play and sport situations. A consideration of culturally-normative gross-motor task demands and its implications for skill learning follows.

As children grow older, the band-width of task demands expected of them in play and games situations changes in an exponential manner. As mentioned earlier, the movement consistency phase of motor development lasts until approximately the age of six years. During this period, the child learns to initiate and control motor skills that respond, in most cases, to relatively low spatial and temporal loads. However, once the child reaches school age, the task demands with which he or she must cope rapidly increase. Bouncing, catching, hitting and kicking balls are skills that are characterized by increases in spatial and temporal uncertainty that demand the use of prediction and other strategic behaviors. Thus, there is an exponential increase in the band-width of task demands from age 6 to 11 years. There is a continued increase in information load up to 15 to 16 years of age. During these years, the "average" child develops cognitive-motor skills that allow him or her to participate in favorite playground games like tetherball and dodgeball or competitive dual and team games such as tennis, hockey, and soccer. Finally, during the later high school and young adulthood years there usually is a decline in task demands and the concomitant skillfulness required in play and games situations as many young people at this age do not participate in competitive team sport activities.

In addition to the changes in task demands with age, there are wide individual differences in skillfulness that may be observed during the developmental period. There is the highly skilled athlete who probably plays for his or her school in a variety of athletic activities. There is the "typical" student who enjoys participating in physical education, intramural sports, and physical recreation activities. The exponential increase in information load during childhood underlines the fact that these children enjoy challenging play situations as evidenced by their voluntary participation in activities that require the handling of much higher information loads. In contrast, there are children who do not acquire sufficient skills to handle the task demands that they meet in many culturally-normative play and sports environments.

In order for school age children to enjoy participating in fast-paced play and sport environments, they must use higher-level cognitive motor plans. As mentioned earlier, the development of these higher level plans depends on practice; not just any practice, but practice which facilitates the development of skillful performance. Unfortunately, from the earliest motor development period, physically awkward children demonstrate a lack of skill in cognitive-motor tasks. As they grow older, this lack of skill becomes a vicious circle, in that crucial practice time within what most children find to be exciting, motivating play situations becomes, in fact, total information overload which results in feelings of failure, frustration, and helplessness. Quite simply, the task

demands that they are expected to respond to in age-appropriate
play and sport situations require skill levels well beyond their
stage of skill acquisition. In effect they are prevented from
practicing. Without practice, these children cannot develop movement consistency let alone movement constancy.

Asking a child reading at a grade one or two level to enjoy
reading a grade six textbook would be similar to assuming that
physically awkward children in grade six can cope with the high
information loads inherent in certain competitive game situations.
Teachers allow for differences in reading abilities; in fact, they
help reading disabled children by using progressions, simplifying
course content, and utilizing specialized techniques. Similar
strategies are needed for physically awkward children if they are to
develop their motor skills.

In summary, the following key observations have been made about
physical awkwardness, skill acquisition, and culturally-normative
performance expectations. First, school-aged children are expected
to participate in physical recreation activities that require considerable proficiency in a wide variety of motor skills. Second,
depending on the nature of the physical activity, be it an informal
game, structured sport, or simply free play, there is a minimal
level of skill proficiency which is required for successful participation. The definition of acceptable proficiency in these motor
skills will vary with the age, sex, and socio-cultural environment
of a person. Third, the development of acceptable motor proficiency
requires consistent practice experiences over long periods of time.
The more difficult the motor skill the more intense and frequent
the practice sessions must be. Fourth, as children grow older, they
are expected to perform perceptually and cognitively-loaded motor
skills in fast-paced game and sport situations if they are to be
accepted by the majority of their peers. Fifth, for a variety of
genetic and environmental reasons, a sizeable number of children do
not develop sufficient proficiency in these culturally-normative
motor skills. These children can be fairly accurately described as
physically awkward. The problems associated with being physically
awkward may seriously affect the personal happiness and general
development of these youngsters. Finally, the above observations
underline the importance of developing appropriate assessment, prescription and remedial services for these children.

The next section of this paper describes the operation of a
motor development clinic for physically awkward children at the
University of Alberta. The clinic is in its formative stages of
development; however, in its operation we have tried to address
questions related to the assessment, prescription, and remedial
services and leisure counselling of these children.

A CLINICAL PROGRAM FOR PHYSICALLY AWKWARD CHILDREN

The Motor Development Clinic offers a number of services to
physically awkward children and their families. The children who
attend the clinic are referred either by their parents or teachers
who are concerned with their general lack of proficiency in motor
skills. After the referral, a child and his or her family may be
involved in some or all of the following services, depending on
their particular needs.

The first service offered by the clinic is a complete gross-
motor assessment that forms part of the information for prescriptive
action. Three different levels of assessment may be used in the
development of a child's motor development profile. Initially, the
child's performance is quantitatively assessed on motor tasks that
include running, throwing, catching, hitting, balancing and agility
skills. A modified battery of tests based on Henderson's revision
(Henderson & Stott, 1977) of the original Stott Test of Motor
Impairment (Stott, Moyes, & Henderson, 1972) is used along with
motor performance items that we have found discriminate unskilled
performers. The child's results on these tests are then compared
with norms that we have developed from relatively large samples of
children in local school systems. In addition, we administer, or
use the available results of the Canada Fitness Awards tests (1980)
to provide a general measure of physical fitness. This initial
assessment is generally completed in a one-hour testing session. If
the results of the assessment indicate that the child does not require further evaluation, the results are shared with the parents
and attendance at the clinic is terminated. However, if a more intensive evaluation is required, the child returns for the next phase
of motor evaluation which requires two more one-hour testing sessions.

The second phase of the motor assessment process is based on
the use of observational checklists that facilitate the rating of
a child's performance on culturally normative motor tasks such as
running, throwing, catching, hitting, kicking, and skipping. The
qualitative evaluation of a skill is simplified through the use of
a series of short phrases that describe the key components that
should be present in the beginning, intermediate, and mature stages
of skill acquisition. Careful observations of the performance of
these skills usually indicate whether the child has difficulty
reacting to the spatial-temporal demands of a task or executing the
response components of the task. The child's performance is
evaluated under increasingly more difficult environmental demands.
For example, in the evaluation of catching skills, the initial
evaluations are based on tasks in which the child stands in a
stationary position to catch a tennis ball thrown from a distance
of approximately ten feet; however more difficult tasks require
the child to catch balls while running across the room at greater

distances or bouncing and catching the ball off a wall within set
time limits. The evaluation of culturally-normative tasks in this
manner seems to more adequately tap the domain of motor skills that
are required for participation in real-life play and sport situations, thus increasing the predictive and ecological validity of
the assessments.

The second service offered by the clinic involves parent interviews regarding the child's physical health, scholastic progress,
general social-emotional development, and current leisure time
interests. During the interview, special emphasis is placed on the
child's past participation in play, games, and sport situations.
An attempt is made to assess the parents' interest in physical
activity and the degree to which they value their child's participation in such activities. These interviews also place the child's
difficulties into a family perspective in which the expectations of
the parents can be more readily understood. Followup interviews
with the parents are usually held to discuss the remedial program
that has been recommended, the progress of the child, and possible
leisure-time activities which the child might enjoy.

Following the motor assessment procedures and the above interviews with the parents, the child's performance scores and background information are summarized on a motor development profile
form. The staff member responsible for the evaluation meets with
the clinic staff who have been involved in the testing to develop
a preliminary program prescription for the child. This prescription
is then presented to the clinic staff which includes senior under-
graduate students in adapted physical education, graduate students,
and staff in the Department of Physical Education. The prescription
is evaluated and usually modified by this group before it is
accepted as a possible plan for remedial action. A parent report
is completed which includes a summary of the motor performance
evaluation and an outline of the recommended followup action. At
this point in the process the child might become involved in the
instructional and/or leisure-counselling programs of the clinic.

The instructional program in the Motor Development Clinic
accepts only ten children at a given time. A senior student is
assigned responsibility for one pupil under the supervision of the
program coordinator. The instructional objectives for the program
are based on the interests of the child, the suggestions of the
parents and/or teacher, and the estimated readiness of the child to
learn selected motor skills. Motor skills that will be used
frequently and in a variety of settings are usually prescribed for
instruction. The instructional program meets twice each week in
one of the university's gymnasiums. Parents agree to bring the
child to the program for a series of at least ten lessons which
meet after the regular school day.

A major premise underlying the instructional program is the
belief that a child must be able to perform the basic response
pattern of a motor skill within low spatial-temporal uncertainty
situations before moving into more demanding ones. Therefore,
instruction on selected skills is based on a graded progression
through a sequence of individual, dual, or small group activities
that facilitate plenty of practice on the skill. If a child has
difficulty performing a basic movement response, individualized
instruction based on task analysed instructional sequences is used
(Watkinson & Wall, 1979). However, if the child has mastered the
rudiments of a response action but has difficulty in situations with
higher spatial-temporal uncertainty, the instructor selects activities for practice that gradually increase in task difficulty. For
example, a child may be able to run for a short distance in a
relatively straight line, but exhibits difficulty in changing
direction, or stopping and starting quickly. In such a situation,
the instructor uses progressive learning experiences that require
the child to practice these latter skills individually in a relatively open space and then increases the difficulty by gradually
adding stationary objects or other children to the practice environment.

Likewise, in catching skills, the instructor uses a variety of
practice activities that require a similar level of skill for successful performance. As the child learns to catch in relatively
easy situations, the instructor increases the task difficulty of the
practice situation. By reducing the size of the ball, increasing
its speed, adding other players, or varying the size of the playing
area, the instructor progressively increases the information load.
The degree to which the instructor increases the task difficulty
depends on how well the child performs the catching skills in the
various practice situations. Initially, the instructors use progressive learning experiences that have been written by clinic
staff; however, they modify these progressions to meet the specific
needs of each child. The purpose of the progressions is to ensure
that the child independently performs as many successful perfor-
mances as possible.

At the conclusion of each instructional program, the child and
parents meet with a member of the staff to discuss the child's
future participation in recreation activities. Prior to this
meeting, the staff member completes a community resources inventory
that summarizes the various recreation programs and activities that
are available in the child's community. The activities that are
identified include not only physical recreation programs but social
and cultural activities as well. Within these broad categories,
individual, dual, team, and family activities are identifed with
special attention being paid to the range of physical recreation
activities that are available in the community. Activities such as

swimming, jogging, skating and cross-country skiing that require
minimal perceptual and cognitive decision-making are identified as
well as the various dual and team sport activities that require
higher levels of skill proficiency for participation.

During the instructional program the staff member encourages
the child to identify the recreation activities that interest him or
her and records them for use in the leisure counselling process.
The staff member also completes a motor performance profile that
summarizes the child's proficiency on key culturally-normative motor
skills.

The staff member uses the above information to match the
physical and social skills of the child, his or her leisure-time
interests, and the degree of support available from the family with
the various recreation opportunities that are available in the
community. A list of suggested recreation activities is discussed
with the parents and the child. The importance of selecting activities in which the child feels comfortable and has a chance of
experiencing success is stressed with the parents. Based on the
leisure counselling discussion, the child, in conjunction with his
or her parents, selects a few recreation activities that seem to
suit his or her needs and capabilities. Information on registering
in programs and monitoring the child's experiences in them is also
given to the parents. They are encouraged to discuss any problems
the child experiences in these activities with the staff member.

The older the child and the greater the discrepancy between
actual and expected performance the more important is the role of
the above leisure counselling process. Educating physically awkward
children and their parents in the wise selection of physical recreation activities may alleviate some of the social difficulties
experienced by these children. If physically awkward children can
be helped through the crucial school years by encouraging them to
select activities that they enjoy they may continue participating
in physical recreation activities throughout their lives.

In summary, the Motor Development Clinic in the Department of
Physical Education at the University of Alberta offers a number of
services to physically awkward children and their parents. First,
when a child registers with the clinic an initial assessment is
completed which includes a developmental and family history,
quantitative and qualitative evaluations of motor performance,
selected fitness measurements, and an inventory of the leisure-time
interests of the child and family. Second, an individual motor
development and activity profile is completed on each child which
summarizes the above information as a basis for the prescription of
remedial strategies. Before prescribing motor skills for instruction, the staff estimate the readiness of the child to learn the

skills and consider the degree to which the child might use them in
play situations. Therefore, only a few key skills are prescribed
for instruction. Third, the instructional program is designed to
teach specific skills through the use of progressive learning experiences that encourage the child to practice the skill in increasingly more difficult environmental situations. The basic premise
of the instructional program is that children learn most effectively
if they independently perform the skill and variations of it in
increasingly more difficult practice situations. Fourth, a leisure
counselling process is used to encourage the children and their
parents to select physical recreation activities that the physically
awkward child will enjoy. Throughout the program a great deal of
emphasis is placed on having fun.

SUMMARY AND CONCLUSIONS

     In contrast to the thrust of recent articles that have dis-
cussed the relationship of physical awkwardness to academic achievement, this chapter has been concerned with the effects of awkwardness within the physical recreation domain. A new definition of
physical awkwardness was proposed along with a discussion of the
following observations of it from a motor development perspective.

     First, the crucial factor in the definition of physical awk-
wardness is the discrepancy between the attained skillfulness of
awkward children and the skillfulness of the majority of their
peers. Second, the level of skillfulness that is expected increases
as the majority of children progress through the general stages of
motor development. Third, the rate of motor skill development
depends on a complex interaction of genetic and environmental
factors; however, an essential factor in the acquisition of motor
skills is progressive practice within increasingly more difficult
environmental situations. Fourth, motor learning is facilitated
when learners can cope with the major task demands in practice
situations. Unfortunately, very often physically awkward children
cannot cope with the task demands of culturally normative practice
situations; therefore, this lack of meaningful practice increases
their skill deficiencies in relation to their peers.

     A careful consideration of the above observations on physical
awkwardness clearly supports the need for sensitive remedial
strategies to help physically awkward children. An example of one
remedial program for physically awkward children was described.
Further theoretical and program development research on the assessment, prescription, instruction, and leisure counselling of physically awkward children is needed. The search for an understanding
of physical awkwardness and the means to alleviate it has only
begun.

References

Bayley, N. A.   The development of motor abilities during the first three years. Monographs of the Society for Social Research in Child Development, 1935, 1(1), 1-26.

Bouchard, C.   Genetics and motor behaviour. In D. M. Landers and R. W. Christina (Eds.), Psychology of motor behaviour and sport. Champaign, Illinois: Human Kinetics, 1976.

Canada Fitness Awards Test. Ottawa: C.A.H.P.E.R., 1980.

Elliott, J., & Connolly, K.   Hierarchical structure in skill development. In K. Connolly and J. Bruner (Eds.), The growth of competence. New York: Academic Press, 1974.

Fitts, P. M., & Posner, M. I.   Human performance. Belmont: Brooks/ Cole, 1967.

Gesell, A., & Amatruda, C. S.   Developmental diagnosis. New York: Hoeber, 1941.

Glencross, D. J.   Psychology and sport. Sydney: McGraw-Hill, 1978.

Gordon, N., & McKinlay, I.   (Eds.), Helping clumsy children. Edinburgh: Churchill Livingstone, 1980.

Gubbay, S. S.   The clumsy child. Philadelphia: Saunders, 1975.

Henderson, S. E., & Stott, D. H.   Finding the clumsy child: genesis of a test of motor impairment. Journal of Human Movement Studies, 1977, 3, 38-48.

Keogh, J. F.   Movement outcomes as conceptual guidelines in the perceptual-motor maze. Journal of Special Education, 1978, 12, 321-330.

Keogh, J. F.   Consistency and constancy in preschool motor development. In H. J. Muller, R. Decher, and F. Schilling (Eds.), Motor behaviour of preschool children. Schorndorff: Hofman, 1975.

Keogh, J. F.   Incidence and severity of awkwardness among regular school boys and educationally subnormal boys. Research Quarterly, 1968, 39, 806-808.

Higgins, J. R.   Human movement: an integrated approach. Saint Louis: Mosby, 1977.

Lindsay, P. L., & Palmer, D.   Playground game characteristics of Brisbane primary school children. Education Research and Development Committee Report No. 28. Canberra: Australian Government Printing, 1981.

Marteniuk, R. G.   Information processing in motor skills. New York: Holt, Rinehart and Winston, 1976.

McDonnell, P. M.   Patterns of eye-hand coordination in the first year of life. Canadian Journal of Psychology, 1979, 33(4), 253-267.

Schmidt, R. A.   Control processes in motor skills. Exercise and Sport Sciences Reviews, 1976, 4, 229-261.

Stott, D. H., Moyes, F. A., & Henderson, S. E.   A test of motor impairment. Slough: N.F.E.R. Publishing, 1972.

Sutton-Smith, B., & Rosenberg, B.   Sixty years of historical
    changes in the game preferences of American children. Journal
    of American Folklore,   1961, 74, 17-46.
Torgeson, J. K.   Characteristics of research on learning
    disabilities. Journal of Learning Disabilities,   1980, 13(9),
    531-535.
Watkinson, E. J., & Wall, A. E.   The PREP play program: play
    skill instruction for young mentally retarded children.
    Edmonton: Alberta Education, 1979.
Welford, A. T.   Fundamentals of skill.   London: Methuen, 1968.
Whiting, H. T. A.   Acquiring ball skill.   London: Bell, 1969.
Whiting, H. T. A., Clarke, T. A., & Morris, P. R. A.   A clinical
    validation of the Stott Test of Motor Impairment. British
    Journal of Social and Clinical Psychology,   1969, 8, 270-274.
Zelazo, P.   From reflexive to instrumental behaviour. In L. P.
    Lipsitt (Ed.),   Developmental psychobiology: the significance
    of infancy.   Toronto: Wiley, 1976.

EPILOGUE

THE NEXT MOVES

Peter Bryant

Department of Experimental Psychology
Oxford University, Oxford, England

Everyone who has tackled the problem of learning difficulties
in children will agree on at least one thing. The problem is a
damnably elusive one. It is not just that it poses the questions
that are usually associated with other childhood disorders - questions about causes, treatment, and prognosis; it also raises some
dilemmas which are peculiarly its own. We cannot even be sure who
these children are, or by what criteria to recognize them. We know
that they are there; we can be certain that their difficulty is
serious and all too common. However, we are not yet sure exactly
how to spot them.

One could suggest a number of possible reasons for this obscu-
rity. An obvious possibility, which happily can be ruled out, is
neglect or lack of interest. These children may have been ignored
in the past, but nowadays they have a secure position in the scene
of academic research, and indeed this book tells the story of the
growing interest in this subject.

Whether the rather large amount of attention which they are now
receiving from research workers is doing them much good is an open
question. Barbara Keogh's chapter shows us very clearly not only
why, to date, the considerable body of research on learning difficulties, viewed as a whole, has not told a coherent story, but also
the reasons why there is still no general agreement about how to
identify these children.

It is to a great extent a matter of the variety of approaches.
The fate of the study of learning difficulties is rather like that
of the noble and beautiful but sadly battered city of Budapest,
which for many centuries always managed to be on the edge of one

271

empire or another. Being located at the point where different
regimes met, meant that it was the focus for many disagreements,
many of them violent.

In much the same way, the study of learning difficulties is
also a meeting point, this time of many different sciences. Educators, speech therapists, pediatricians, psychologists and psychia-
trists alike have tried to explain and to suggest ways of alleviating the problem of learning difficulties. Since the people in
these disciplines have radically different approaches, speak very
different languages and often have rather different goals, it is not
at all surprising that there should be the kind of divergences which
Barbara Keogh's work has shown to exist.

It is confusing, but it is almost certainly to the good. For
it means that the disciplines are meeting. The key to a successful
solution to the problem of learning difficulties must lie in successful coordination of the many types of knowledge involved, and a
lot of the present confusion is simply what happens when people who
are very different from each other try to cooperate.

The least that can be claimed for this book, and the conference
on which it is based, is that it represents a step towards this sort
of coordination. The conference involved the fields of Special Education, Physical Education and Developmental Psychology, and in the
end the people who attended had learned a great deal about other
approaches.

More than this, it soon became clear that a number of central
questions were being posed again and again. They were couched in
different terms and came from very different points of view. But
the questions were the same, and there was often agreement about how
they could be answered. To pose the right questions is surely half
the battle, and this book's achievement is possibly to reach a surprising degree of agreement about what the right questions might be.
Let us turn to them.

THE PROBLEM OF HETEROGENEITY

Children with learning difficulties fail to make progress at
school, though on all other reasonable indices such as intelligence
or strength of motivation they should be faring reasonably well.
That much is certain. However, this group has another quality which
is almost as certain, and that is that it is not a homogeneous one.
The exact nature of difficulties varies from child to child, and it
follows that the causes of their different difficulties probably
vary as well. This fact is not so welcome because it poses great
problems for research workers and educators: for the research

worker because he has to face the daunting task of splitting up the
major group into justifiable subgroups and for the educator because
these different subgroups may well need different types of intervention. The research workers' problem is probably the primary one,
simply because the question of different types of teaching cannot be
properly answered until the subgroups themselves are established.
How then to establish them?

Different ways are reported in this book. Some recommend a
longitudinal approach using statistical techniques which allow different groups to emerge without the help (or hindrance) of prior
hypotheses. Others, such as Torgesen, adopt a virtually opposite
strategy. Torgesen's hypotheses are his starting point and his
evidence is not longitudinal. Yet it is interesting to note that
many of the conclusions are similar despite the radical differences
in methods. This surprising degree of agreement illustrates an
important point. The problem of heterogeneity will, in the end,
probably be solved by heterogeneity in approaches. I mean that we
need a variety of methods, and not just the two that have been
mentioned.

Differences in educational outcome should also be a source of
valuable information. Some children have difficulties with reading
and spelling and mathematics as well, and others just with reading
and spelling. Still others fail in spelling even though they read
quite normally for their age (Frith, 1980). There are a few whose
difficulties concentrate around mathematics without much accompanying failure with written language. What differences underly these
variations and how are they related to the subgroups which emerge in
studies like those of Torgesen?

This is not all. Another potent source of hypotheses about
subgroups must be the intense case study (Bradley, Hulme, & Bryant,
1979). Individual case studies have not featured explicitly very
largely in this book, but they surely lie behind many of the hypotheses in it. No scheme is viable unless, eventually, every single
individual child with evident learning difficulties can be slotted
into one of its categories with ease, and it is probably pointless
to concoct such a scheme without first going into reports of individuals in some detail. Hagin, Beecher, and Silver's use of the
case history in this volume, combined with traditional diagnostic
criteria, is a good example of the value of paying detailed attention to particular cases.

Hagin, Beecher, and Silver also make another point; it is that
the form of children's difficulties almost certainly changes with
age, so that the same child may show a different pattern of symptoms
at different ages. This complicates the picture even further, but
cannot be ignored, and it means that we must invest more in longitudinal studies.

Another requirement of any such scheme must be to explain the connections between different symptoms. There are, for example, some learning disabled children who cannot read, and who also have poor verbal memories. However, we should not just demonstrate this relation. We must explain exactly why and how poor verbal memory might lead to educational difficulties. This leads us to the second main question which concerns causes.

CAUSAL SEQUENCES

Everything in the end must turn on causes and causal sequences. We cannot attempt to put a condition right with any degree of confidence, unless we know what has provoked it.

Let us take a simple example. Many children with learning difficulties are clumsy (Gordon & McKinlay, 1980; Gubbay, 1975; Hulme, 1981), though many are not. What role, one must ask, do these motor difficulties, suffered by some learning disabled children, play in their eventual failure in school? There are at least two possibilities.

One, which accords with at least two influential theories about child development (Held, 1965; Piaget, 1954), is that the children's motor difficulties might directly cause their educational problems. These theories hold that the child constructs his knowledge about the spatial properties of his environment (and eventually, in Piaget's view, about the rules of logic), through his interactions with the environment. Such interactions are at first largely determined by the movements they make, and the sequence of developmental changes during the early years of life is said to depend a great deal on changes in the child's motor abilities. Both approaches could be said to involve versions of the idea of the image of the act, which is the central theme of the chapter in this book by Whiting and den Brinker. The act is the important thing, and in the first few years of life these acts are movements. If these views are correct, the general intellectual development of a child whose movements are poorly controlled would be at peril. It is a short step from there to ask whether the root cause of the atrocious spelling and mathematical disorganization which are found in so many clumsy children with learning difficulties are the product of their poor motor development.

It is not of course the only possibility. The obvious alternative is that the two things - clumsiness and learning difficulties - have nothing directly to do with each other, though they may both be directly caused by some other root abnormality.

The distinction is important, because it has clear practical implications. If the first alternative is right, then you must throw all your resources into correcting the motor problem and compensating for the poor motor development: educational success should follow as day follows night. But if the second alternative is the correct one, then motor remediation is no longer the central concern. It should help of course to alleviate a problem which is distressing in its own right, but it will have little effect, one way or the other, on the child's spelling or on his sums.

This suggests an obvious test. There are indeed plenty of motor programs which have been tried often enough. It has, in fact, been the common view for some time that these programs, however successful in alleviating children's motor difficulties, have no consistent effect on their educational levels at all - a pattern of results which suggests that the clumsy children's learning difficulties are not the product of their poor motor development. However, one of the most striking features of Jack Keogh's cogent account of this research is that these studies may not have been a sufficient test of the hypothesis. His novel and important point that these programs tended to use movement in teaching rather than to teach movement, means in effect that the hypothesis still awaits a test.

How good a test of a causal hypothesis is a training experiment or an interventional program? I think that, on its own, it has its weaknesses. Suppose, for example, that one got a negative result, i.e. that motor programs which genuinely were motor programs still had no effect. It is easy to see that nothing much could be concluded from this result because it could be that the wrong methods were used. A training program which improved movement, but left the child's educational prowess untouched, might have failed on the educational side because its methods were too artificial and bypassed the kinds of experience which normally go with successful motor development. It is a problem of artificiality.

If training experiments have their problems what about other methods? Among the other techniques used to track down causes, the major one is correlation and, in the field of learning difficulties as we shall see, the concealed correlation. One might as well start with the weaknesses of the simple correlation since these are now very widely recognized. On its own it is ambiguous, because the direction of cause and effect is ambiguous. For example, parents who talk to their children a lot tend to have children who read well (Whiteman & Deutsch, 1968). But this could as well be because interested, lively children who learn to read soon are also the sort of children who provoke their parents into interesting conversations, rather than the alternative of the parent affecting the child. Either could affect the other, and as Barbara Keogh pointed

out, there is also the very plausible transactional view that both
in addition could change the other.

These possibilities are so familiar nowadays that the simple
correlation rarely plays a part any more in serious discussions of
causes of learning difficulties, and there does not seem to be any
reason to shed any tears over its demise. Indeed, the issue would
hardly be worth mentioning, were it not for the fact that studies
which carry all the faults of simple correlations and which are
actually concealed correlations, unfortunately, do play a very large
and uncriticized part in the discussions of learning difficulties,
and particularly of difficulties with reading.

I refer to cross-sectional experiments which compare two groups.
One is usually a group of backward readers whose IQ's are normal but
whose reading ages fall well below their actual ages. The other is
a normal control group with children of the same age and IQ, but
with normal reading ages. Thus, the only difference between the two
groups is in their reading skill, and the usual procedure is for the
experimenter to test both on some aspect of behavior, usually linguistic or perceptual, which he thinks has led to the first group's
reading failure.

When it is found that the backward readers are worse on whatever ability is at issue, it is usually concluded that a deficit
which causes that reading failure has been found. Until recently
(Bradley & Bryant, 1978) the opposite possibility, that the deficit
might actually be caused by the backward readers' relative lack of
successful reading experiences, received scant attention. Yet here
is the same problem again: this kind of experiment is as ambiguous
about causal directions as the simple correlation. The only difference is that the danger was recognized earlier and more widely with
the correlation. The cross-sectional equal age/equal mental age
experimental comparison between backward and normal readers still
forms the bulk of evidence that exists on reading difficulties, and
causal inferences from this type of evidence are very rarely criticized. They ought to be.

So the three main methods which have been used to work out
causal sequences are, when used on their own, quite fallible. But
we need not despair because there is a solution to each problem.

Correlations are more convincing when they are no longer
simple. As has been long recognized, and as the Lesgold and Resnick
chapter amply demonstrates, they become much more powerful if they
are repeated at different points in time. If A causes B, A should
precede B in time and this would mean that A at time 1 should correlate more highly with B at time 2 than vice versa. The logic behind
such causal arguments based on these comparisons is set out so

clearly by Lesgold and Resnick that it needs no repeating here. The question is, how strongly do these correlations over time indicate causes. The answer, in the end, must be that though they are a great deal better than the simple correlation, they are not completely convincing on their own. If A at time 1 does turn out to correlate with B at time 2 and not vice versa this could be because they are both controlled by some third factor (e.g. maturation) which has its effects on A before it influences B.

However, this problem is soluble, and again Lesgold and Resnick have shown us how. You must combine the longitudinal data with training experience. If the correlations over time suggest that A causes B, then look to see what effect improving A eventually has on B. We have seen that on their own, training experiments run the risk of being arbitrary and artificial. But if they are linked to the naturally occurring events which have been plotted by correlations, their problem disappears. We need longitudinal studies combined with intervention to study the causes of learning difficulties. They have practical impact and they do track down causes. So far there have been few studies of this sort. Let us hope for more.

STRATEGIES

There was a time when the universal view of cognitive development was that it was simply a matter of abilities being acquired during childhood. The young child, it was generally held, started off without this and that ability, and had to acquire it one way or another as he grew older. The developmental psychologist's job was to describe when and how he managed to fill in these holes in this intellectual armour.

This view has received a severe jolt in recent years because it is clear that many of the intellectual abilities thought to be absent until late in childhood turn out to be present and in good order at very early ages (Bryant, 1974; Donaldson, 1978; Stone, Smith, & Murphy, 1974). As a result, it has become increasingly clear as Das and Mulcahy's, Barclay and Hagen's, and Downing's chapters show, that there is another important question to be asked: How do children organize abilities which they do possess so that they apply them at the appropriate time? This change of emphasis is of great importance to the study of learning difficulties. Under the old view, the only course one could possibly take was to look for missing abilities. Which particular mechanism or ability did children with learning difficulties lack? Now one can take seriously the possibility that these children, who are typically poorly organized, suffer not from a defect of this or that ability, but from a failure to deploy their abilities properly at the right time.

There is no need to repeat the accumulating evidence, already
described in the three chapters just mentioned, that many of the
problems in learning difficulties might be due to a certain disorganization in deploying strategies rather than to any outright defect.
Two things remain to be said about this new evidence. The first is
that it is a message of hope, since it's surely going to be an
easier matter to show a learning disabled child to deploy a strategy
which he already possesses than to try to instill in him a totally
new ability.

The second point is that the issue has raised once again the
question of consciousness. This question, which is a difficult one,
has always loomed large in the discussion of motor abilities.
Whiting and den Brinker's 'image of the act' is an example of a
theory which gives conscious awareness an important role in skilled
movements. But there are other hypotheses, such as Bruner's and
Connolly's, which, though not necessarily contradictory, place far
greater emphasis on skilled movements no longer needing conscious
control.

Now the same difference of emphasis and even of opinion can be
found in studies of reading and of reading failure. Lesgold and
Resnick lay great emphasis on the importance of 'automaticity' in
word recognition. In other words, they think that it is important
that the skill becomes so well ingrained that it ceases to depend on
conscious processes. Downing on the other hand, and many others
elsewhere, stress that conscious awareness of phonological and semantic phenomena is a necessary precursor of skilled reading.

Phonological codes have received the most attention from those
interested in the importance of conscious awareness in learning to
read (Bruce, 1964; Gleitman & Rozin, 1977; Lundberg, 1978; Marcel,
1980). The usual argument is that children of five or six come to
school for the first time with a fully fledged but implicit phonological code. They can produce, string together, understand and
distinguish all the phonemes when they speak and listen to speech
with no difficulty at all. But they are unaware of doing so: they
do not explicitly realize that the words and syllables which they
speak and hear can be broken up into phonemic segments. They have
an implicit but not a conscious knowledge of phonemes, and they have
to learn to be consciously aware of speech segments before they can
cope with the alphabetic code which also deals with these segments.

So deployment depends on access, according to this argument,
and access might be particularly difficult for children with learning difficulties. It is a coherent hypothesis but it needs to be
closely questioned. As Chukovsky (1963) has shown and Lynette
Bradley has recently confirmed in a large scale study of 400 pre-
readers, most young children are interested in and can detect rhymes

and alliteration at quite young ages. Rhymes and alliteration crop
up naturally in early word play, and they involve speech segments
very explicitly.

Furthermore, much of the evidence which seems to suggest that
young children do not appreciate phonemes is itself quite questionable. Bruce (1964), for example, showed that young children could
not subtract a phoneme from a word (What would 'stand' sound like
if you took away the 't'?), but it is quite possible that children
could be aware of phonemes without being able to take them notion-
ally away. The Liberman/Shankweiler tapping test (Liberman,
Shankweiler, Liberman, Fowler, & Fischer, 1977) in which young
children learn to tap out syllables much more easily than phonemes
is just as tenuous. Tapping is a rhythmic task and the rhythm of
words is caught in their syllables, not in their phonemes.

So the possibility arises that children might be aware of
speech segments but sometimes do not apply this knowledge properly
when they have to learn to cope with written words. Or, it is
possible that normal readers are always aware of the phonological
structure of words, while backward readers, who have been shown to
be particularly bad at detecting rhymes (Bradley & Bryant, 1978),
are not.

We do not know yet, but the general point can still be made
that if a child does not deploy a psychological strategy, which is
somewhere in his intellectual repertoire, when he should, this does
not mean that he is quite unaware of this strategy. Consciousness
and deployment are not the same thing, and since consciousness is
such a tricky notion it would perhaps be more profitable for us to
concentrate on deployment first.

CONCLUSION

In my chapter I have made three pleas, all of which have been
taken from the chapters of this book. The first is that we should
accept that a lot of what Hagen rightly called "the fuzziness" of
the concept of learning difficulties is caused by it being something which interests several different disciplines. Indeed we
should welcome it because it is probable that cooperation between
different approaches should lead to a genuinely new system.

The second plea is for more longitudinal work. Of course
longitudinal studies are more difficult to arrange and more expensive than the much commoner cross-sectional studies, but, for two
reasons, they are also more fruitful, at any rate as far as learning
difficulties are concerned. The first reason is that the longitudinal method is the only way to sort out the problem of heterogeneity,
by showing how much of the variation between children is due to

underlying differences between subgroups and how much to changes in
the pattern of difficulties which occur with age. The second point
about longitudinal studies is that, combined with intervention, they
are the only really effective way of tracking down causal sequences.
Unless we do manage to pin these sequences down, we will have little
either of theoretical or of practical value to say about learning
difficulties.

The third plea is that those studying the form of learning difficulties should take the more recent insights of developmental
psychology seriously, and study the question of the deployment of
abilities as well as the more traditional question of the possession
(or not) of these abilities. We should recognize that the child's
own considerable abilities may be masked by an unusual degree of
disorganization.

These seemed to me to be the main points of the conference on
learning disabilities. I have never seen them made so clearly, or
seen them made together, before this conference. It should have an
effect.

References

Bradley, L., & Bryant, P. E.   Difficulties in auditory organization
     as a possible cause of reading backwardness. Nature, 1978, 271,
     746-747.
Bradley, L., Hulme, C., & Bryant, P. E.   The connection between
     different verbal difficulties in a backward reader: a case study.
     Developmental Medicine and Child Neurology, 1979, 21, 790-795.
Bryant, P. E.   Perception and understanding in young child.   New
     York: Basic Books, 1974.
Bruce, D. J.   An analysis of word sounds by young children.
     British Journal of Educational Psychology, 1964, 34, 158-170.
Chukovsky, K.   From two to five.   Cambridge: Cambridge University
     Press, 1963.
Donaldson, M.   Children's minds.   Glasgow: Fontana, 1978.
Frith, U.   Cognitive processes in spelling.   London: Academic
     Press, 1980.
Gleitman, L. R., & Rozin, P.   The structure and acquisition of
     reading 1: The relations between orthographies and the structure
     of language.   In A. Reber and D. L. Scarborough (Eds.),   Toward
     a psychology of reading.   Hillsdale, New Jersey: Lawrence Erlbaum
     Associates, 1977.
Gordon, N., & McKinlay, I.   Helping clumsy children.   Edinburgh:
     Churchill Livingstone, 1980.
Gubbay, S. S.   The clumsy child.   London: Saunders, 1975.
Held, R.   Plasticity in motor systems.   Scientific America, 1965,
     213.

Hulme, C.   Reading retardation and multi-sensory teaching.   London:
    Routledge and Kegan Paul, 1981.

Liberman, I. Y., Shankweiler, D., Liberman, A. M., Fowler, C., &
    Fischer, F. W.   Phonetic segmentation and recoding in the
    beginning reader.   In A. Reber and D. L. Scarborough (Eds.),
    Toward a psychology of reading.   Hillsdale, New Jersey: Lawrence
    Erlbaum Associates, 1977.

Lundberg, I.   Aspects of linguistic awareness related to reading.
    In A. Sinclair, R. J. Jarvella, and W. J. M. Levelt (Eds.), The
    child's conception of language.   Berlin: Springer-Verlag, 1978.

Marcel, T.   Phonological awareness and phonological representation.
    In U. Frith (Ed.), Cognitive processes in spelling.   London:
    Academic Press, 1980.

Piaget, J.   The construction of reality in the child.   New York:
    Basic Books, 1954.

Stone, L. J., Smith, M. T., & Murphy, L. B.   The competent infant.
    London: Tavistock, 1974.

Whiteman, M., & Deutsch, M.   Social disadvantage as related to
    intellective and language development.   In M. Deutsch, I. Katz,
    and A. R. Jensen (Eds.), Social class, race, and psychological
    development.   New York: Holt, Rinehart and Winston, Inc., 1968.

CONTRIBUTORS

BARCLAY, Craig, R.   Department of Psychology, University of
    Michigan, Ann Arbor, Michigan, U.S.A.

BEECHER, Ronnie   Department of Psychiatry, New York University
    School of Medicine, New York, U.S.A.

BRADLEY, Lynette   Department of Experimental Psychology, Oxford
    University, Oxford, England.

BRYANT, Peter, E.   Department of Experimental Psychology, Oxford
    University, Oxford, England.

DAS, J. P.   Centre for the Study of Mental Retardation, University
    of Alberta, Edmonton, Alberta, Canada.

den BRINKER, B.   Department of Psychology, Interfaculty of Physical
    Education, The Free University, Amsterdam, The Netherlands.

DOWNING, John   Faculty of Education, University of Victoria,
    Victoria, British Columbia, Canada.

HAGEN, John, W.   Department of Psychology, University of Michigan,
    Ann Arbor, Michigan, U.S.A.

HAGIN, Rosa, A.   Fordham University School of Education and New
    York University School of Medicine, New York, U.S.A.

KEOGH, Barbara, K.   Department of Education, University of
    California, Los Angeles, California, U.S.A.

KEOGH, Jack, F.   Department of Kinesiology, University of
    California, Los Angeles, California, U.S.A.

KNIGHTS, Robert, M.   Department of Psychology, Carleton University,
    Ottawa, Ontario, Canada.

283

LEONG, Che, K.   Institute of Child Guidance and Development,
    University of Saskatchewan, Saskatoon, Saskatchewan, Canada.

LESGOLD, Alan, M.   Learning Research and Development Center,
    University of Pittsburgh, Pittsburgh, Pennsylvania, U.S.A.

MULCAHY, Robert, F.   Department of Educational Psychology,
    University of Alberta, Edmonton, Alberta, Canada.

RESNICK, Lauren, B.   Learning Research and Development Center,
    University of Pittsburgh, Pittsburgh, Pennsylvania, U.S.A.

SILVER, Archie, A.   University of South Florida Medical School at
    Tampa, Tampa, Florida, U.S.A.

SNART, Fern, D.   Department of Educational Psychology, University
    of Alberta, Edmonton, Alberta, Canada.

TORGESEN, Joseph, K.   Florida State University, Tallahassee,
    Florida, U.S.A.

WALL, A. E.   Department of Physical Education, University of
    Alberta, Edmonton, Alberta, Canada.

WHITING, H. T. A.   Department of Psychology, Interfaculty of
    Physical Education, The Free University, Amsterdam, The
    Netherlands.

DATE DUE

OCT 13 1988			
GAYLORD			PRINTED IN U.S.A.